Three Days

Before The Sun

© Gaylena Andrushko

"God said, 'Let there be light!'
...and there was light."
Genesis 1:3

Warren LeRoi Johns

Author of Chasing *Infinity, Time Zero, Genesis File, Beyond Forever, Ride to Glory,* and *Dateline Sunday, U.S.A.*

Three Days Before the Sun

Copyright © Warren LeRoi Johns, 2011
All rights reserved. No part of this book may be used or
reproduced, except brief quotes or reviews, without written permission.

© Historic Old Colorado, John Herbert Johns unpublished portfolio

**John Herbert "Jay" Johns camera lens captured
century-old scenes of Colorado life from the "mile high city."
Selections from his private porfolio illustrate portions of this book.**

Cover
Rebecca Paschal's stunning design features **Mike Norton's**
glimpse of the sun from Monument Valley's perspective.

Publisher: www.GenesisFile.com
Printed in the United States of America
Lightning Source, LaVergne, TN

ISBN: 978-0-615-45670-6
Library of Congress Control Number: 2011911645

Dedicated to the worship of
The Infinite, Eternal God of the Universe,
Author of true science and the
Creator of all life!

"The exquisite precision of science, revealed in the "Book of Nature," is God's gift!
More than mutually compatible, science and Scriputre function inextricably commingled
in synchronous harmony, refuting the 'flaws' and 'holes' in the chance hypotheses."
WLJ

A Fresh Look at the Evidence

I---Chance on Trial---7
Intelligent Design

II---Fact-free "Science"---16
Fantasy on Steroids

III---Cosmic Convergence---27
Life Friendly Ecosystem

IV---Superstitious Nonsense—42
Spontaneous Generation

V---Paging Sherlock Holmes---54
Missing Links

VI---*"According To Its Kind"*---64
Stasis

VII---Imagination Run Amok---73
Living Cell

VIII---*"Nonsense of a High Order"*---85
DNA, Language of Life

IX---*"Refuse Material of Nature's Workshop"*---99
Mutations

X---Gregor Mendel's Monkey Wrench---111
Sand in the Gears

XI--- Tree Roots---121
The Plant Kingdom's Discontinuity

XII---The Laughing Mouse---130
Irreducible Complexity

XIII---An Inconvenient Truth---142
Extrapolation

XIV ---Playing the Time Card---155
Radiometric Dating

XV---*"Let there be light..."* ---166
Timeliness

XVI---Time Zero---177
Young Life, Old Earth

XVII---Life's Slippery Slope---189
Devolution

XVIII---Mother Earth's Facelift---201
Global Gully-Washer

XIX ---Tracking The Perfect Cataclysm---210
Fossil Residue

XX---Man Hunt---223
Human Genealogy

XXI---Command Center---237
Human Brain

XXII---Seven Billion Miracles---248
It's About Us

XXIII---The Truth About God---260
It's About Him

XXIV---God's Greatest Gift---268
Life:---*"Original, Unborrowed, Underived"*

Endnotes---281

Sources---299

Author---311

Light of the World

A few thousand years before the present, the first day of creation week arrived three days before there would be a sun and a Solar System!

A transcendent light enshrouded what had been water-covered, formless matter, floating in cosmic darkness for unknown eons of time. The burst of light announced the presence of the infinite Creator, who, in seven literal days, introduced life with a supporting ecosystem, transforming the inert, sterile blob of chemical elements into our vibrant,"Blue Planet."

The brilliant light enshrouding earth that first day of creation week, announced the presence of *"The Spirit of God,"* overwhelming darkness by delivering the Creator's shining brightness, without aid of the sun.

There was a reason for the time-lapse before the sun's creation!

Awed humans tend to worship readily visible power. While the sun sustains and nourishes life, it is not the source of life. The sun is man's servant, not a master or god to be worshipped.

Three days later, after *The Spirit of God* bathed earth in brightness, and after the atmosphere, dry land, and the Plant Kingdom were in place, *"God made two great lights---the greater light to govern the day and the lesser light to govern the night…And there was evening, and there was morning---the fourth day."*

Taking the Genesis account to mean what it says, the inert chunk of water-covered land, floating in cosmic darkness, pre-existed the creation of life on earth. Creation week featured the miraculous introduction of young life on old matter, a few thousand years before the present.

Any any fossil remnant of a former life form, discovered encased in earth's hardened sediment, doesn't take on the deep time age of that burial matter anymore than a family pet, buried on the "back forty," takes on the ancient age of its grave site.

The ball of fire that rules the Solar System, radiating a steady stream of light rays at a speed of 186,282 miles-per-second, emitting energetic power beyond easy comprehension, is a servant of life on earth---not its god.

Our worship is due the infinite **God of the Universe**, our **Creator**---the source of all power, light and life.

He is now and forever the *"light of the world."*

I

Chance on Trial

Intelligent Design

"It is reasonable for historical scientists to infer that an intelligence acted somehow in the past to produce the existing information-rich sequences in living cells." [1]

Stephen C. Meyer

Historic Old Colorado, John Herbert Johns unpublished portfolio

This quaint, rustic bridge spanning a stream at Idledale in the Colorado Rockies reflects design information carried in the blueprint of its designer.

Evolution's *Chance Hypothesis* Doesn't Explain the Source of Genetic Information.

Paper and ink alone lack much meaning until an intelligent source inscribes letters and words to create a message. A book with an attractive cover that binds a package of blank white paper would never top the best-seller list. It's the printed message that tells the story and sells the book.

Matter, without information, lacks much meaning, too. Matter, alone, is abstract. The suggestion that the first living cell created itself accidentally, from non-living matter by *spontaneous generation*, epitomizes intellectual incoherence.

Lacking valid scientific explanation for life's beginning on earth, Charles Darwin proposed the faith-based assumption that life originated by accident in some "warm little pond." This was the soil he selected to plant an imagined "tree of life."

Building the chance hypothesis on unscientific nonsense makes no more sense than trying to make a map of our world while relying on a flat earth concept of it.

Unaware of the existence of a cell's nucleus, much less its DNA, Darwin and nineteenth century contemporaries like German evolutionist, Ernst Haeckel, erroneously viewed living cells as "homogeneous and structureless globules of protoplasm." [2]

Then came French chemist and microbiologist Louis Pasteur, perhaps best known for pasteurization of milk and his advocacy of vaccination. With his experiments on microorganisms in a swan-neck flask, this scientific genius blew away the superstition of *spontaneous generation*.

Pasteur believed: "Chance favors only the prepared mind," [3] or in the words of the legendary sports manager Branch Rickey, "Luck is the residue of design."

Nor was Pasteur a fan of life by accident. Recalling his findings, he predicted in 1864 that "never will the doctrine of spontaneous generation recover from the mortal blow struck by this simple experiment."[3]

Chemical evolution of original life from non-life has never been demonstrated in a laboratory. Lack of a scientific explanation for the source of genetic information packed in the simplest living cell further compounds the dilemma, gnawing at the core of the chance hypothesis.

Louis Pasteur (1822-1895), brilliant French scientist, provided laboratory proof that *spontaneous generation* was fiction.

The living cell's world was turned upside down in 1931, when German physicist Ernst Ruska introduced a prototype electron microscope. Once perfected, the invention revolutionized life sciences, challenging the chance hypothesis with devastating implications.

Former misconceptions about the cell started to become obsolete. The discovery opened access doors to the cell's secrets, and molecular biology emerged into the science spotlight. Rather than preoccupation with bone fragments unearthed in fossil graveyards, molecular biology exposed the cell's complexity.

Armed with the electron microscope and other sophisticated scientific tools unknown to Darwin, Stanley Miller and Harold Urey attempted to do in a science lab what the chance hypothesis had credited nature with doing accidentally: create a living cell from non-living matter.

In 1953, the ambitious team reported failure.

That same year, two molecular biologists, American James D. Watson and Englishman Francis H. Crick, co-discovered the double-helix molecular structure of DNA, the cell's information storage center. Impressed by the complexity, Crick saw more than "globules of protoplasm."

"An honest man," he wrote, "armed with all the knowledge available to us now, could only state that in some sense, the origin of life appears at the moment to be almost a miracle, so many are the conditions which would have had to have been satisfied to get it going." [4]

DNA stores the genetic information in every cell: nucleotides identified as A (adenine), T (thymine), C (cytosine), and G (guanine). In the view of Bill Gates, "DNA is like a computer program but far more advanced than any software we've ever created." [5]

DNA "gives instructions to the rest of the cell to make proteins, and it passes the same information on to the next generation." More than a random conglomerate of data, DNA's double helix displays an *order*, *shape* and *sequence* unique to each life kind. "Without DNA, living organisms cannot survive." [6]

Thirty-three years after Crick and Watson's discovery earned them a shared Nobel Prize for physiology and medicine, Australian microbiologist and physician, Michael Denton, offered these eye-opening comments:

"...The DNA molecule may be the one and only perfect solution to the twin problems of information storage and duplication for self-replicating automata...

"It is astonishing...that this remarkable piece of machinery which possesses the ultimate capacity to construct every living thing that ever existed on Earth, from a giant redwood to the human brain, can construct all its own components in a matter of minutes and weigh less than 10^{-16} grams.

This "...is of the order of several thousand million-million times smaller than the smallest piece of functional machinery ever constructed by man." [7]

On the brink of the twenty first century, the stage was set for objective life scientists to challenge establishment paradigms and to introduce the role of intelligence as essential to the design of a living cell vested with the ability to store, manufacture and process genetic information. Founded in 1990, Seattle's Discovery Institute attracted scientists committed to reversing the "materialist worldview," while being careful not to speculate on the source of the cell's information. The following year, Attorney Philip E. Johnson's *Darwin On Trial* introduced persuasive evidence that identified many glaring shortfalls in the chance hypothesis.

In 1996, Lehigh University biochemist, Michael Behe introduced to the public the concept of *irreducible complexity* in living systems; he also suggested that *intelligent design* accounted for the remarkable complexity of the cell. Two years later, William Dembski, a cutting-edge mathematician with a University of Chicago doctorate, wrote convincingly of the *Design Inference* in nature.

Microbiologist Jonathan Wells, with PhDs from the University of California at Berkeley and from Yale, released his *Icons of Evolution* in 2000, followed by his *Darwinism and Intelligent Design* in 2006. Wells summarized the role of information in the origin equation.

"...Intelligent agents do produce large amounts of information, and since all known natural processes do not (or cannot), we can infer design as the best explnation of the origin of information in the cell." [8]

Stephen C. Meyer's 2009 definitive *Signature of the Cell* put *Intelligent Design* arguments squarely at odds with the chance hypothesis exactly 150 years after Darwin published *Origin of Species.* [9]

Meyers wrote: "The 'creation of new information is habitually associated with conscious activity.'" [10]

Less than a century after the electron microscope debuted as an indispensable scientific tool, the chance hypothesis still lingers, despite its lack of any convincing naturalistic explanation identifying the source of the cell's genetic information. Luck-of-the-draw rhetoric rings hollow in the context of verifiable science. The complexity of the "simple" cell continues to astound. Its axiomatic: The first living cell's DNA base pair could not exist much less reproduce without the language of life's imprinted information. A single cell's DNA carries information sufficient to fill 3,000 sets of printed encyclopedias.

"...The capacity of DNA to store information vastly exceeds that of any other known system; it is so efficient that all the information needed to specify an organism as complex as man weighs less than a few thousand millionths of a gram...Each gene is a sequence of DNA about one thousand nucleotides long."[11]

Cell proteins derive from an array of twenty basic amino acids. The odds against the 2,000 enzymes [catalysts from proteins] essential to the simplest life form appearing spontaneously from inorganic matter, at one time and in one place, runs in the range of $10^{40,000}$:1.[12] Such an event stretches astronomically beyond the 10^{50} impossible range.

"...Since most proteins are, on average, 300 amino acids long, the DNA needed to make just one protein would have to be approximately 900 letters long." [13]

"The simplest extant cell, Mycoplasma genitalium---a tiny bacterium that inhabits the human urinary tract---requires 'only' 482 proteins to perform its necessary functions and 562,000 bases of DNA (just under 1,200 base pairs per gene) to assemble those proteins." [14]

So where did those estimated 562,000 bases of DNA, pre-loaded in *Mycoplasma genitalium,* come from? The accident conjectured by the chance hypothesis, or design from an intelligent source?

Nucleotides A, T, C and G represent the genetic code *letters* that combine to form the *words* of instruction which in turn combine to form the genes or *chapters*, which then compose the chromosomes or *books* of the complete *library* of genetic information. Each living cell of a plant or an animal requires a complete *library* of genetic information in order to function. The more complex the life form is, the more extensive the genetic information *library*.

Consider the overwhelming mathematical odds against one single protein in one microscopic cell having its information library of books, chapters, words, and letters arranged accidentally in proper *order*, *shape* and *sequence*.

The genome of a mammal contains a string of "from two to four million" symbols "...that would fill two thousand volumes---enough to take up a library shelf the length of a football field. All this is in the tiny chromosomes of each cell." [15]

As for humans, "each human body contains a galaxy of cells---100 trillion---."[16] It has been calculated that if "...all the copies of the DNA in all the cells of your body were straightened and laid end to end they would be about 50 billion kilometers long,"[15] the line would stretch from the earth to beyond the solar system.

Even if time were capable of creating information, thirteen or fourteen billion years is not enough for the chance hypothesis to deliver genetic information to a single living cell--much less to an intelligent, fully-functioning human system.

"The odds of getting even one functional protein of modest length (150 amino acids) by chance from a prebiotic soup is no better than 1 chance in 10^{164}...With odds standing at 1 chance in 10^{164}...the probability is 84 orders of magnitude (or powers of ten) smaller than the probability of finding the marked particle in the whole universe.

"Another way to say that is the probability of finding a functional protein by chance alone is a trillion, trillion, trillion, trillion, trillion, trillion,

trillion times smaller than the odds of finding a single specified particle among all the particles in the universe." [17]

Stymied by the "far greater problem" in explaining how first life created itself from non-living matter, the chance hypothesis brushes aside the obvious *design inference* in favor of accidental "magic." Side-stepping any viable rationale explaining the debut of first life, the hypothesis embraces a, faith-based secular theology, fabricated from obsolete assumptions. Chemical evolution suffers from head-in-the sand mentality that appears tone deaf to molecular biology's persuasive new discoveries.

Sir Isaac Newton (1643-1727) anchored the Scientific Revolution with his definition of the three laws of motion and his understanding of universal gravitation. Newton's giant intellect recognized God as the Creator of science as revealed in the "Book of Nature." Newton's laws symbolize the pristine beauty of natural science beckoning mankind to open nature's book, turn the pages and pull back the curtain of discovery. Nature's book carries the signature of the Author of Science. Humans did not create science; rather, they are positioned to scratch its surface.

Just as the living cell dictates patterns of plant and animal life, every scientific discipline is vested with precise information, the kind for instance that enabled humans to explore the moon. Rocket scientist Werner von Braun recognized "the existence of a superior rationality behind the existence of the universe."

Science and religion don't represent an either/or choice but exist as a mutually complimentary package. Scientific discovery inevitably widens the reality gap between verifiable knowledge and the unproven vagaries of the chance hypothesis.

Someone once pointed to man's thumb and forefinger as independent components of the hand but united at the base so that, together, they can clasp an object. Life's origin and genetic science is best observed when science and religion, acting jointly as thumb and index finger, embrace truth in the discovery process.

Faith is essential to either the chance hypothesis or belief in the Master Designer. Evolution, with its commitment to life's chance beginning in dark chaos and ending in black nothingness, requires faith in the abstract. Creationists assert that the information and design inherent in all plant and animal life offer convincing evidence of the Master Designer's touch.

Scientists such as Behe, Dembksi, Wells and Meyer challenge the chance hypothesis. They have built the case for *Intelligent Design* using valid scientific evidence. Still, ID's design imperative makes no attempt to delve into theology or to identify the source of genetic information.

Three Days Before the Sun respects the persuasive quality of the evidence supporting the *design inference*, but then takes a faith-based step: It offers evidence that the Author of all science and the source of all information imperative for life is God, the Creator of all things.

II

Fact-free "Science"

Fantasy on Steroids

"I believe that one day the Darwinian myth

will be ranked the greatest deceit

in the history of science."[1]

Søren Løvtrup

Historic Old Colorado, John Herbert Johns unpublished portfolio

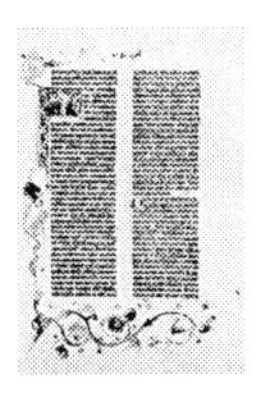

Darwin imagined a self-created cell would evolve gradually, eventually replacing itself by increasingly complex life forms. Living and fossil evidence refute this dream.

Charles Darwin rejected the Genesis narrative of the creation week miracle and the source of light that turned darkness into day. Undeterred by lack of verifiable scientific evidence explaining the origin of first life, and relying on his own "warm little pond" scenario--Darwin plunged ahead with his grand scheme alleging every diversified plant and animal life form shared common ancestry with some unidentified original, simple cell. In his initial burst of fervor to illustrate this extravagant notion, Darwin postulated that given time, a bear might evolve into a marine mammal "as monstrous as a whale." This land animal-to-ocean-going-whale fiction appeared as a figment of Darwin's fertile imagination in the first edition of his *Origin of Species*.

"In North America the black bear was seen by Hearne swimming for hours with widely open mouth, thus catching, like a whale, insects in the water.

"Even in so extreme a case as this, if the supply of insects were constant, and if better adapted competitors did not already exist in the country, I can see no difficulty in a race of bears being rendered, by natural selection, more and more aquatic in their structure and habits, with larger and larger mouths, till a creature was produced as monstrous as a whale." [2]

This abstract, thumb-your-nose rationale conflicts head-on with the biblical reference that "God created great whales." [3]

The nineteenth century naturalist never enjoyed the privilege of donning hip-waders and fraternizing with beluga whales in a 55° Chicago water tank designed to entertain the human species.

Darwin's whale-of-a-tale scenario imagining a bear that evolved into a sea-going whale exposed the glaring inconsistency of a flawed idea!

©Audrey Snider-Bell

© janprchal

These aquatic monsters weigh-in at a ton or more, and in captivity open their mouths to be fed, have their tongues petted, and nuzzle noses with humans who fork over $200 for the privilege.

Does anyone seriously believe the beluga that bedazzles humans evolved coincidentally and gradually from some ferocious, land-based, hairy-bear ancestor?

Evolution's fact-free fantasy--from fish-to amphibians-to-reptiles-to-mammals—posed a conundrum: Where does the whale fit in? If the warm-blooded whale, a mammal, descended from an egg-hatched, cold-blooded reptile, the chain of organic life transitions stood in jeopardy, seriously out-of-kilter. A winter hibernating, fury, land-based mammal with paws and claws, transiting to a sleek denizen of the deep-blue-sea "as monstrous as a whale" ignores logic and lacks supporting evidence.

Some have suggested that *Ambulocetus*--an extinct, 11 foot, four-legged, land-based critter, lacking a blowhole, flippers and missing a whale's rudder-like tail-- is a "walking whale"? This idea matches Darwin's bear-to-whale absurdity.

Why would descendants of a self-sufficient, land based bear, seek refuge in the ocean's salt-water depths in order to survive? Had the imagined bear-to-whale evolution actually occurred, why do bears roam the landscape after Darwin predicted that "we may safely infer that not one living species will transmit its unaltered likeness to a distant futurity." [4]

Understandably, Darwin himself presumably recognized the preposterous absurdity: He abandoned the conjecture in all post-1859 editions of *The Origin of Species.*

Any bear attempting transit to whale status by natural selection or otherwise would be destined for a watery grave--as helpless as a car without wheels or as impractical as a desk-top printer without an ink cartridge.

While this ambitious bear-to-whale postulation never made it past 1859's first edition of *The Origin of Species*, Darwin later underscored his vision of the make-believe with a conjectured molecule-to-man scenario resulting from a series of un-designed natural accidents.

Charles Darwin lived in a cocoon of economic privilege, never having to earn a living with his hands.

Darwin entered the world in 1809, near the peak of a power pyramid, at a moment when "Rule Britannica" symbolized empire. British flagged slave ships plied the Atlantic, while wealth was built on the backs of foreign nationals and an exploited home grown labor force that lived at the whim of the owners of the mills and mines. By some strange irony, Abraham Lincoln, author of the "Emancipation Proclamation," shared a February 12, 1809, birth date with Darwin, propagator of racial supremacy.

The well-to-do Darwin family supported young Charles' university studies while he explored careers in medicine and in the ministry. The family connection earned him crew membership as a naturalist on the HMS

Beagle. Later, married to a Wedgewood fortune heir, he settled in a country estate near London.

While tin mines were closing in Cornwall and men were dying in the Crimean War, he enjoyed the comfort of an estate managed by an eight-person servant staff. Thriving on his personal pedestal of privilege, he made no apology for inherited wealth and gratuitous social perks.

"The presence of...well-instructed men, who have not to labour [sic] for their daily bread, is important to a degree which cannot be overestimated; as all high intellectual work is carried on by them, and on such work material progress of all kinds mainly depends." [5]

His words indicate awareness that he lived near the apex of the not-so-subtle class structure saddling European cultures.

"Without the accumulation of capital the arts could not progress; and it is chiefly through their power that the civilized races have extended, and are now everywhere extending, their range, so as to take the place of the lower races." [5]

He appears to have correlated accumulation of wealth as a valid measure of human worth. With personal vision warped by nineteenth-century superstitions, Darwin fretted that vaccination would lead "to the degeneration of a domestic race." Darwin's survival of the fittest theme inspired his taking an ill-conceived shot at vaccination, believing it "preserved thousands, who from a weak constitution would formerly have succumbed to small-pox...the weak members of civilized societies propagate their kind...this must be highly injurious to the race of man...a want of care, or care wrongly directed, leads to the degeneration of a domestic race." [6]

Could this outspoken stance have been a left-handed slap at the discoveries of Louis Pasteur, a courageous and pioneering advocate of vaccination? Or was Darwin oblivious to Pasteur's findings?

Regardless, in the classic tradition of his English heritage, he maintaind a stiff upper lip while pledging to "bear without complaining the undoubtedly bad effects of the weak surviving and propagating their kind." [5]

Given the flourishing slave trade sanctioned by British society, racism tainted his public pronouncements. He awarded pinnacle status to his own social environment.

"Various races differ much from each other…the capacity of the lungs, the form and capacity of the skull…in their intellectual, faculties." [7]

"The western nations of Europe…immeasurably surpass their former savage progenitors and stand at the summit of civilization…" [8]

He predicted, thanks to evolution's touted "progress toward perfection," the time would come "…not very far distant, as measured by centuries, the civilised [civilized] races of man will almost certainly exterminate and replace throughout the world the savage races." [9]

England's Queen Victoria, a Darwin contemporary, occupied the British throne for 63 years, beginning in 1837. Without a tip of the hat to the reigning monarch, Darwin revealed a gender bias suggesting that after moving past monkey ancestry, "Man has ultimately become superior to woman." [10]

To the Queen's credit, Darwin wasn't banished to the Tower of London for saluting male chauvinism. Undeterred, the naturalist elaborated further:

"If two lists were made of the most eminent men and women in poetry, painting, sculpture, music, comprising composition and performance, history, science, and philosophy, with half-a-dozen names under each subject, the two lists would not bear comparison.

"We may also infer…that if men are capable of decided eminence over women in many subjects, the average standard of mental power in man must be above that of a woman." [11]

Charles took his cue and shaped his biased views from personal research and the leanings of his grandfather. Erasmus Darwin speculated that "all warm-blooded animals have arisen from one living filament" that could continue, "to improve by its own inherent activity" with the improvements passed along to descendants. [13]

In 1837, just back from a five-year odyssey aboard the HMS Beagle, the 28-year-old Charles sketched a "tree of life," depicting branches of new and different life forms sprouting, by chance, from some unknown original cell.

Twenty-two years later, his "tree of life" blossomed into the ponderous *Origin of Species* treatise depicting the constant shift of unstable life forms, from the simple to the complex, en route to an unpredictable "biologic transit stop."

While sweeping under the rug the "far higher problem" of just how life first created itself, accidentally, from non-living matter, he embraced "natural selection" as the critical means for culling out and preserving the fittest.

With phraseology more than faintly reminiscent of grandfather Erasmus' musings, Darwin's grand scheme imagined a taxonomic tree with new and different branches sprouting from the accrual of a continuous series of miniscule increments of genetic change, transiting to new and distinctly different organisms.

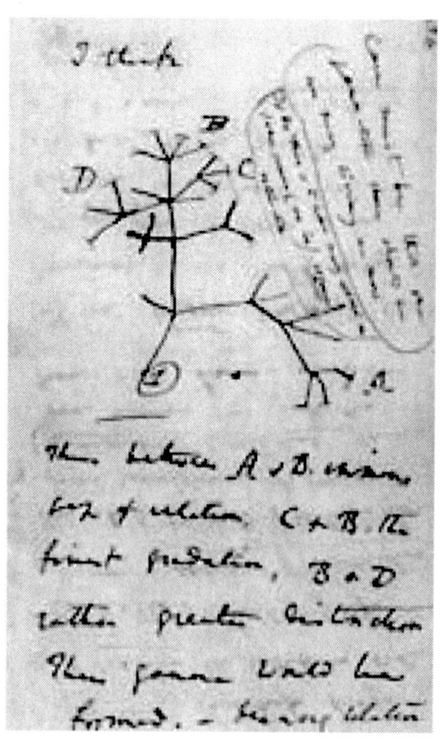

A 28-year-old Darwin designed this "I think" tree in 1837.

Genetically preposterous bear-to-whale mismatches expose the chance hypothesis' extravagant conjectures while doing nothing to validate Darwin's vision of a chain of organic life transiting upward from that first living cell *species*, and on to *genus*, to *family*, to *order*, to *class*, to *phylum*, and eventually to *kingdom*.

Despite Darwin's intense plant research, the junction of the plant and animal kingdoms remains vague. In fact, the inconvenient truth stands resolute: No evidence exists suggesting real change within a genome has ever jumped genetic limits by evolving radically different descendants that creep gradually up the *family-order-class-phylum-kingdom* taxonomic ladder.

Crafting a concept on unproven assumptions,
Darwin recognized his novel formula required mega-chunks
of deep time in order to allow natural selection to work its "magic."

Consider Darwin's words: Natural selection can only take "advantage of slight successive variations; she can never take a great and sudden leap, but must advance by short and sure, though slow, steps." [14] Those "slight successive variations" supposedly result in "the production of new forms" that will cause "the extinction of about the same number of old forms." [15]

No one has seen a partially evolved organ, such as an eye, lung, heart or brain produced by "slight successive variations." In the pointed vernacular of down-home Texans, "You can put your boots in the oven but that can't make them biscuits."

Natural selection is a genuine fact of nature. But don't hold your breath expecting Darwin style evolutionary change if "natural selection" relies either on acquired physical characteristics or mutations to jump genetic barriers inherent in every genomic kind.

Just one example: A female's sexual selection, expressing preference for a mate, occurs throughout animal kingdom kinds. Males attract a female's attention by performing exotic dances, feats of strength and issuing "romantic" sounds, unique to a species.

It's the female's prerogative to select the winner.

Rather than evolving some new and different organism, her selection works to preserve, protect and maximize the genetic quality of her own species---the opposite of the Darwinian postulate.

Working at cross-purposes to evolution theory, natural selection acts counter to the evolution equation by screening out harmful mutations.

"Natural selection can serve only to 'weed out' those mutations that are harmful, at best preserving the 'status quo.' " [16] It "...can act only on those biologic properties that already exist; it cannot create properties in order to meet adaptational needs." [17]

"No one has ever produced a species by mechanisms of natural selection." [18]

Is sexual selection as an obvious demonstration of natural selection in action? Of course. Does it then follow that natural selection is a tool of tortured trail of biologic transition projected by Darwin? Hardly.

Darwin imagined that a simple form of first life might have emerged, accidentally, from some vaguely defined, "warm little pond." [19] Then given enough time, that simple life supposedly evolved into "an aquatic animal ...with the two sexes united in the same individual;" then on to "some fish-like animal" ancestral to a "reptile-like or some amphibian-like creature;" ultimately morphing into an "ancient marsupial" and finally "higher mammals." [20]

"All the higher mammals are probably derived from an ancient marsupial, and this through a long line of diversified forms, either from some reptile-like or some amphibian-like creature, and this again from some fish-like animal." [20]

Darwin assured skeptics: "Man is descended from a hairy quadruped, furnished with a tail and pointed ears, probably arboreal in its habits..." [19] and that "early progenitors of man were no doubt once covered with hair, both sexes having beards; their ears were pointed and capable of movement; and their bodies were provided with a tail..." [21]

He alleged humans traced their ancestry to "Old World monkeys." [22] His ponderous manuscripts appear less certain as to just where grass, vines, bushes and trees fit into evolution's master scheme.

Although the master of his Downe estate never retracted the core of his imaginings, he did express misgivings to trusted confidents.

Michael Denton, a biochemist writing late in the twentieth century, saw "imagination" as an inadequate substitute for "evidence."

"If anyone was chasing a phantom or retreating from empiricism it was surely Darwin, who himself freely admitted that he had absolutely no hard empirical evidence that any of the major evolutionary transformations he proposed had ever actually occurred.

"It was Darwin, the evolutionist, who admitted in a letter to Asa Gray, that one's 'imagination must fill up the very wide blanks.' " [23]

Without bowing to the Genesis narrative of life's origin but lacking any rational alternative, Darwin, aware of the shallow depth of the "warm little pond" scenario, conceded original life may have been "originally breathed by the Creator into a few forms or into one." [24]

In a burst of candor, Darwin recognized the feeble fabric of his idea, acknowledging, "I am conscious that my speculations run beyond the bounds of true science." [25] He admitted that his theory seemed to be "a mere rag of an hypothesis with as many flaw[s] and holes as sound parts." [26]

Before his 1882 death, the English naturalist verbalized uneasy equivocation in a whiff of prescience, confessing doubts about his thesis. He fretted he may "...have devoted my life to a phantasy [sic]." [27]

His concerns were justified. But considering the realities of his 19th century environment, the case can be made to cut the guy some slack. Laboratory tools were primitive; he knew nothing about mutations, DNA or even a cell's nucleus; he lived in a cultural cocoon that touted empire and scorned non-European races; and the established religion he knew survived by government subsidy rather than by a believer's faith.

Sophisticated 21st century minds can't claim these defenses. The plethora of scientific discoveries only widen the gap between the chance hypothesis and evidence confirming earth's intelligent design.

To qualify as science, "explanations of large classes of phenomena must make testable predictions and be falsifiable...there must be a way to make an observation that could disprove the explanation." [28]

Mislabeling a sow's ear doesn't make it a silk purse. Hyping evolution's abstract assumptions doesn't qualify unproven imaginings as science.

Facts have a way of fading in the hands of artful *spinmeisters* capable of whitewashing assumptions, affixing catchy labels, and wrapping conjecture in cosmetic mantles of pseudo-authenticity.

The human mind's quest for knowledge is staggered by the complexity of the cosmos and earth's ecosystem. Faith in an infinite but unseen Supreme Being who designed and created what surrounds us boggles finite comprehension. But to reject faith in a living, infinite Intelligence in favor of life that created itself from non-life demands a stretch of faith that abandons the essence of the very rationality that atheists and doubters claim to embrace. The brightest finite minds shrink to miniscule nothingness in the presence of the Creator of the universe.

Some devout evolutionists accuse creationists of rejecting science. This charge rings hollow. How could those who believe God authored science be blind to the facts of its fundamental principles?

Pioneer scientists like Isaac Newton and Louis Pasteur held personal faith in God. Creationists don't reject science but do challenge evolution's unproven assumptions and abstract assertions paraded as science.

Evolution, Darwin style, never happened. The theory propagates faux science and dark philosophy. It starts in unknown blackness and ends in eternal death.

To paraphrase Winston Churchill, building an idea on a bubble is as futile as "a man standing in a bucket trying to lift himself up by the handle."

III

Cosmic Convergence

Life-friendly Ecosystem

"You may find it hard to believe God could make everything from nothing. But the alternative is that nothing turned itself into everything." [1]

Mark Cahill

© Paul Prescott

The moon's orbit can be calculated hundreds of years, past or future. Evolution doesn't explain how life "evolved" on our Blue Planet but bypassed the sterile moon.

"The visible universe contains about 70,000 million-million-million stars---that's a 7 followed by 22 zeros."[2]

Earth's life-friendly ecosystem could not have created itself. Against all odds, some twenty-first century theorists came to the rescue, proposing an unusual "good" news/"bad" news scenario. The "good" news: The unconfirmed report that the mystery of the origin of first life on Planet Earth has been solved. The "bad" news: The suggestion is that our ultimate ancestor may well have been a "biochemical moron."

The report starts simply enough, announcing that "…many scientists believe that viruses evolved very early on, possibly even earlier than everything else. If so, they are not merely some ornamentation on the tree of life but rather may compose its very roots." [3]

Our first ancestor was a virus parasite? Could a parasite virus live without access to a prior life form to latch onto?

"We humans…are nobody's great idea; we are the fortunate mistakes of countless biochemical morons. That's evolution. It is humbling but somehow comforting." [3]

Comforting? Finding some great-grand-pappy virus smirking at us smugly from the pages of the family's ancestry album?

Taking a swipe at Intelligent Design theory, the report asserts, "…the viruses appear to present a creation story of their own: a stirring, topsy-turvy, and decidedly unintelligent design where life arose more by reckless accident than original intent, through an accumulation of genetic accounting errors committed by hordes of mindless microscopic replication machines." [3]

"Unintelligent" may be the understatement of the millennium.

Throw in "reckless accident…mindless…genetic accounting errors… mistakes…" and, of course, those "countless biochemical morons"and the imagined trip to antiquity begins to resemble a journey to la-la land.

While the verbiage may inspire a field day of punditry, the idea carries a serious side. Recognizing viruses to be older and more complex than once believed and that possibly they may compose the "very roots" of the "tree of life" is an idea that represents eye-rolling poppycock.

Where's the substantiating evidence? And where's any verifiable explanation as to the source of genetic information for the virus? Formulation of a living cell, capable of reproducing itself, has never been created in the laboratory much less accidentally generated spontaneously in nature. Not even a parasite virus is capable of independent living; it has to latch onto an already living host.

Big Bang proponents speculate that all matter and every star in the universe resulted from an exploding dot no bigger than a period at the end of a sentence.

How can "something" explode from "nothing?" What caused "nothing" to explode into "everything?" If "nothing" was all that existed, what could go "bang?" Could there be absolutely "nothing"---no energy, no matter, no light, not even empty space, or no space, in a universe without boundaries? Do the laws of physics support the notion that every piece of matter floating in a 14-billion light year observable universe could have been compacted in a dot no bigger than the period closing a sentence?

And where and how did that all originate?

If energy and matter can neither be created nor destroyed, then where did the energy and matter compressed within that miniscule dot originate before the supposed inflation from the Big Bang? And how can a destructive force create a cosmic universe of something out of nothing?

Verifiable answers to these queries ride waves of silence.

For the cosmos to have created itself, as is now conjectured, the basic laws of physics, as now understood, would have to be suspended, modified or abandoned.

"...When you squeeze the entire universe into an infinitesimally small, but stupendously dense package, at a certain point, our laws of physics simply break down. They just don't make sense anymore." [4]

Astrophysicist Peter Coles from England's University of Nottingham throws cold water on the not-so-hot idea.

"There is little direct evidence that inflation actually took place...It is a beautiful idea that fits snugly with standard cosmology...but that doesn't necessarily make it true.

"We don't know for sure if inflation happened...In a way we are still as confused as ever about how the universe began." [5]

So much for the alleged Big Bang.

No evidence confirms life exists on neighboring planets Venus, Mars, or Saturn, so what cosmic magic engineered life on earth?

Why life here and not on the moon?

Big bangers look to an explosive force, supposedly occurring several billions of years before the present, as having self-created multi-trillions of bits and pieces of matter, which formed orbiting spheres, parading in space in some good-luck sequence of universal order.

Other academics attribute the harnessing of all the factors essential to create a life-friendly environment to a Master Designer!

Thanks to predictable order in space, mathematical measurements plot the time and place of cosmic orbits with uncanny precision, whether the time dimension dips millenniums into the past or extends outside the reach of an uncharted future.

Explosions observed on earth rip matter apart, leaving fragments of ruble, strewn helter-skelter in disorganized trash heaps.

Where is evidence a cosmic explosion created order out of chaos?

How could a cosmic explosion create the mathematical balance evident in galaxies---from nothingness?

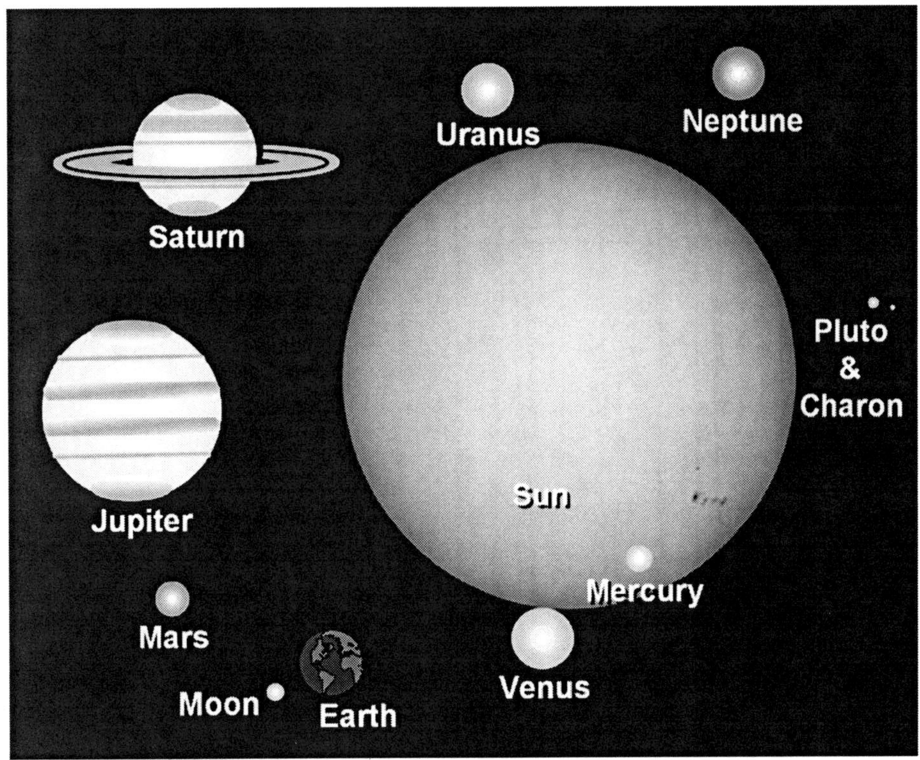

www.webcrawler.com

Suspended in space, without cables or foundations, earth moves in three directions simultaneously---spinning on its axis; orbiting the sun; and floating in sync with the other components of the solar system within the 100,000 light year diameter of the Milky Way galaxy.

Nothing random about this system of planets spinning about the sun, nothing chaotic, and it does not seem like the product of chance. The most brilliant minds can't duplicate this cosmic balancing act. Keeping free-floating spheres in perpetual motion, floating in repetitious orbits, without strings—the idea confounds our best minds.

That's just the beginning of flabbergasting realities.

The length of a day on our Blue Planet is 24 hours, with 365.256 days to a year. Venus rotates once every 243 earth days. A Venus solar day runs 116.75 earth days. Every planet in the solar system moves at its own speed; the Solar System displays a precise symphony of mathematical balance.

Planetary orbits display variances of distance and shapes with tendencies ranging from the round to the elliptical. Axis tilt angles are unique for each planet. Most rotate counter clockwise to the east while Venus, Uranus and Pluto rotate clockwise to the west.

Does this sound like fall-out from an explosion?

Can a theoretical explosion explain the origin of environmental conditions essential for life's genesis?

"No matter how large the environment one considers, life cannot have had a random beginning…" [4] Heavy doses of intellectual flimflam can't disguise reality!

"There are about two thousand enzymes, and the chance of obtaining them all in a random trial is only one part in $(10^{30})^{2000} = 10^{40,000}$, an outrageously small probability that could not be faced even if the whole universe consisted of organic soup...

"If one is not prejudiced either by social beliefs or by a scientific training into the conviction that life originated on the Earth, this simple calculation wipes the idea entirely out of court…" [6]

"…The difficulties in producing a protein from the mythical prebiotic soup are very large, but more difficult still is the probability of random processes producing the simplest living cell…" [7]

Think about it.

"The concept that all the parts of the first living thing preexisted, and its formation was simply a matter of spontaneous generation there from is mathematical absurdity, not probability. All present approaches to the problem of the origin of life are either irrelevant or lead to a blind alley." [8]

Bradley and Thaxton reason that "…even assuming that all the carbon on earth existed in the form of amino acids and react at the greatest possible rate of 10^{12}/s for one billion years…the mathematically impossible probability for the formation of one functional protein would be $\sim 10^{-65}$." [9]

Sir Fred Hoyle calculated that "the likelihood of even one very simple enzyme arising at the right time in the right place was only one chance in 10^{20} or 1 in 100,000,000,000,000,000,000… that about 2,000 enzymes were needed with each one performing a specific task to form a single bacterium like E. coli." [10]

Hoyle, with his colleague Chandra Wickersham, estimated that "the probability of all of these different enzymes forming in one place at one time to produce a single bacterium, at 1 in $10^{40,000}$.

"This number is so vast that it amounts to total impossibility." [10]

While sticking to his life-by-accident postulate, Darwin tossed a sop to skeptics, conceding life might have been "originally breathed by the Creator into a few forms or into one..." [11]

Hardly an endorsement of the biblical creation week, the following allusion from the last paragraph of *Origin* may have appeased some Christians looking for compromise, while giving root to his "one" or a "few" simple life forms supposedly growing a "tree of life."

"It is often said that all the conditions for the first production of a living organism are now present, which could ever have been present.

"But if we could conceive in some warm little pond, with all sorts of ammonia and phosphoric salts, light, heat, electricity, etc., present, that a protein compound was chemically formed ready to undergo still more complex changes, at the present day would be instantly devoured or absorbed, which would not have been the case before living creatures were formed." [12]

Without that first-ever living cell making its début, Darwin's *Origin of Species* dream of evolving branches on a "tree of life" would be stranded rootless with no chance to grow---even in several billion years.

Darwin's biological evolution hypothesis can't fly without the initial launch from that chemical "warn little pond" scenario which posits the alleged action of energy sources forming organic compounds in the atmosphere "washed down by rain and accumulated in the primitive oceans until they reached the consistency of a hot dilute soup. According to this model, life appeared from the chemical reactions and transformations that took place in this prebiotic soup." [13]

Big problem here: evidence of the magic elixir is a no-show.

"Prebiotic chemical soup, presumably a worldwide phenomenon, left no known trace in the geological record." [14]

So-called "dawn rocks" from Western Greenland, conventionally dated at 3.9 billion years before the present and reputedly the oldest known dated rocks on the planet, show nothing resembling prebiotic soup.

"Rocks of great antiquity have been examined...and in none of them has any trace of abiotically produced organic compounds been found...

"Considering the way the prebiotic soup is referred to in so many discussions of the origin of life as an already established reality, it comes as something of a shock to realize that there is absolutely no positive evidence for its existence." [15]

Hubert Yockey dismisses "primeval soup" as a non-event.

"The origin of life by chance in a primeval soup is impossible in probability in the same way that a perpetual motion machine is impossible in probability." [16]

Abiogenesis didn't happen but should be viewed as "just a relic of the cosmology of the time it was invented...

"There is no evidence that a 'hot dilute soup' ever existed. In spite of this fact, adherents of this paradigm think it ought to have existed for philosophical or ideological reasons...Scientists are divided into segregated schools that do not even agree on the standards of scientific inquiry..." [17]

With respect to the "prebiotic soup theory of the origin of life...objective scientific principle of a search for the truth is replaced by the subjective aesthetic principle of a well-constructed story." [17]

The never-happened spontaneous generation of life from chemical non-life comes burdened in the collateral fiction of the never-was *Ur-schleim*, an imaginary slime-like material existing deep in the ocean, allegedly the nursery for first life in a "self-origination" format.

"In science one must follow the results of experiments and mathematics and not one's faith, religion, philosophy or ideology. The primeval soup is unobservable since, by the paradigm it was destroyed by the organisms from which it presumably emerged." [18]

Darwin contemporary, Sir William Dawson, didn't think much of evolution theory.

He labeled it "one of the strangest phenomena of humanity…a system destitute of any shadow of proof, and supported merely by vague analogies and figures of speech….

"Now no one pretends that they rest on facts actually observed…Let the reader take up either of Darwin's great books, or Spencer's 'Biology,' and merely ask himself as he reads each paragraph, 'What is assumed here and what is proved?' and he will find the whole fabric melt away like a vision…

"Evolution as an hypothesis has no basis in experience or in scientific fact, and… its imagined series of transmutations has breaks which cannot be filled." [19]

Some evolutionists, kick the first-life can down the road by suggesting life survived the rigors of outer space travel and rode to earth after originating somewhere else in cosmic space following the hypothetical Big Bang. The question left hanging: "How did that first life create itself?"

The likelihood of natural forces delivering the ingredients essential to sustain life, simultaneously in one place, at one instant in time, and without intelligent input is as preposterous as the discredited notion of spontaneous generation.

Microscopic particles, one ten-millionth the mass of an electron, are believed to be the most miniscule form of matter known to man. The atom, with its positively charged nucleus encircled by an array of electrons, ranks as the smallest unit of the elements charted in the periodic table displayed in high schools. Combinations of inorganic molecular matter can be built from a mix of these elements in carefully calculated, chemical recipes, capable of replication. Molecular bonding, essential to life, requires the presence of no less than 40 different elements. Success is contingent upon electromagnetism functioning within a balanced electron-to-proton mass ratio.

Starting with a formulated mix of oxygen and hydrogen, water flows, a crucial ingredient for life's recipe. Next add a dash of carbon and a touch of sulfur. Finally, bolster the formula with some nitrogen and phosphorous. That's only the beginning of an inorganic base essential for intelligent life.

And don't forget, DNA's pre-programmed information looms as an imperative for cell function and reproduction.

The odds of an explosion, creating an environment friendly to the production of organic life is less likely than all earth's citizens solving a Rubik cube puzzle, simultaneously, in less than a minute---then repeating the exercise, without error, a million consecutive times.

In the solar system, earth alone exhibits a confluence of cosmic coincidences essential to sustain organic life. Plants and animals can't survive except as co-dependent components of an ecological package.

For starters, earth's life-friendly ecosystem thrives on balanced land/water ratios, a reliable supply of liquid water, all nestled within a thin atmospheric envelope with delicately matched ratios of oxygen, carbon dioxide, ozone, and nitrogen. Temperate season are affected by the moon's size and distance from earth. A narrow habitable zone is assured by solar radiation neither too near nor to far from that zone designed for life

The mass, color, location, and luminosity of stars; the inclination of earth's orbit and axis tilt; a terrestrial crust with moving tectonic plates; gravity that keeps feet planted securely; and magnetic fields combine to suggest "if masses did not attract each other, there would be no planets or stars, and once again it seems that life would be impossible." [21]

All this convergence by chance in a split second of time, at a location no more than a micro-mini speck in space, complete with a convenient and readily available array of that array of life-essential elements.

So what's the probability of finding a free-floating space station offering an environment capable of generating and sustaining organic life by chance?

Guillermo Gonzales and Jay W. Richards, analyzed earth's "grand scheme" in their landmark treatise, *The Privileged Planet,* and suggest the chance of all critical environmental factors essential for life appearing simultaneously, seems mathematically off-scale---like winning some cosmic lottery.

This series of conditions friendly to life, all within the cross hairs of one small dot in infinity, beckons recognition of something beyond the luck of the draw. Logic teams with science and Scripture, pointing to a Master Designer.

"It seems as though somebody has fine-tuned nature's numbers to make the Universe ...The Impression of design is overwhelming...We are truly meant to be here." [20]

Recent estimates account for more than 1,000 planets in the cosmos---and still counting. With 10,000 billion billion stars in space, it should be no surprise that life comparable or superior to *Homo sapiens* could exist on other planets with their own co-dependent ecosystems.

Just as life, as we know it, does not generate spontaneously, it has yet to be observed elsewhere in the cosmos, even if odds suggest some form of intelligent life exists elsewhere in the universe.

Explorers from the Blue Planet home base survive in space's hostile environment by carrying their own life-support systems. Fragile human life hinges on access to oxygen, water and food. Air deprivation guarantees suffocation in short minutes. Without water, a person can last several days. Too much water drowns its victims. Without water, death by dehydration awaits. Starvation takes longer but is just as certain. Take away nutritious food, and the most robust person might survive but a few, short weeks. Temperatures too hot or too cold accelerate the death process. Too much, too little, too far, too near, too late, too soon—any factor out-of-kilter and life on earth could not exist. Move the sun 5% closer, and the earth would be scorched. Modify earth's orbit 20% farther from the sun and life would drift into deep freeze.

A transparent atmosphere enables scientific analysis of the heavens. And don't forget that ozone mantle, wrapped protectively around earth, shielding life from overpowering ultraviolet radiation.

Consistent doses of sunlight, radiating beams of ultraviolet and infrared, sustain life. If the sun's relationship to the electromagnetic spectrum shifted imperceptibly, the chance of life could vanish.

Instead of blistering heat or deadly radiation, energy from the sun comes calibrated in a range maximizing a life-friendly environment. Sunlight radiates a collateral bonus in the riot of colors embellishing environments.

Restful sky-blues, backlighting forests of multi-hued greens define the landscape. A rainbow of kaleidoscopic accents, ranging from pastel shades of shimmering pinks and lavenders to crimson-golds, trace the arc of a daily rising and setting sun. Rather than a life in dull black and white, shifting combinations of the sun's rays conspire to induce psychological peace.

Despite somber predictions that the raging inferno at the heart of the solar system is certain to destroy itself someday in a blazing conflagration, the sun ignores dire predictions and keeps "rollin' round heaven all day."

"Almost everything about the basic structure of the universe...the fundamental laws...of physics and the initial distribution of matter and energy...is balanced on a razor's edge for life to occur." [22]

Oxidation renders spontaneous generation of life impossible. Free oxygen inclines to oxidize organic compounds, destroying the chemical building blocks of life. Atoms and molecules tend to bond with oxygen atoms while free oxygen inclines to oxidize organic compounds, destroying the chemical building blocks of life.

"Oxygen was likely present in the early earth's atmosphere." [23]

There is "...strong evidence that oxygen was present on the earth from the earliest ages... Significant levels of oxygen would have been necessary to produce ozone which would shield the earth from levels of ultraviolet radiation lethal to biological life." [24]

If the "early atmosphere was oxygen-free...then there would have been no protective ozone layer. Any DNA and RNA bonds would be destroyed by UV radiation... Either way, oxygen is a major problem." [25]

"...All experiments simulating the atmosphere of the early earth have eliminated molecular oxygen...Oxygen acts as a poison preventing the chemical reactions that produce organic compounds...If any chemical compounds did form, they would be quickly destroyed by oxygen reacting with them..." [26]

"Even if oxygen was not present in the early earth's atmosphere, the absence of oxygen would present obstacles to the formation of life. Oxygen is required for the ozone layer which protects the surface of the earth from deadly ultraviolet radiation. Without oxygen this radiation would break down organic compounds as soon as they formed." [27]

Microbiologist Michael Denton reasons that "in an oxygen-free scenario, the ultraviolet flux reaching the earth's surface might be more than sufficient to break down organic compounds as quickly as they were produced...In the presence of oxygen, any organic compounds formed on the early Earth would be rapidly oxidized and degraded...

"The level of ultraviolet radiation penetrating a primeval oxygen-free atmosphere would quite likely have been lethal to any proto-organism possessing a genetic apparatus remotely resembling that of modern organisms.

"What we have then is a 'Catch 22' situation...If we have oxygen, we have no organic compounds but if we don't have oxygen we have none either." [28]

Ita classic lose/lose scenario---life could never evolve in an atmosphere with oxygen; but once formed, life could not survive without oxygen.

Life's origin and survival also demands water---a rare commodity in the solar system. It is axiomatic: no water, no carbon-based life. How could an explosive big bang in space wrap earth in a wet blanket but leave the circling moon and planets as high and dry as the desert sands?

Genesis describes an originally empty and formless earth with darkness "over the surface of the deep and the Spirit of God was hovering over the waters." [29] Viewed from space, our Blue Planet reflects a bright blue hue. Supposedly 70% of the earth is covered with water. So much so that if the earth's land crust was flattened, water hundreds of feet deep would cover its surface.

Scholars draw fascinating conclusions: 96.5% of earth's water consists of ocean marine water and 0.97% brackish water (Its suggested 95% of the total fossil record consists of marine life). That leaves 2.53% of our entire water supply as fresh water. This comparatively minuscule amount is allocated as: 69.6% glaciers and permanent snow; 30.1% ground water; 0.29% lakes, marshes, and swamps; 0.05% soil moisture; 0.04% atmosphere; 0.006% rivers; and 0.003% living organisms. [30]

"Either life was created on the earth by the will of a being outside the grasp of scientific understanding, or it evolved on the planet spontaneously, through chemical reactions occurring in nonliving matter lying on the surface of the earth.

"The first theory…is a statement of faith in the power of a Supreme Being…The second theory is also an act of faith…assuming that the scientific view of the origin of life is correct, without having concrete evidence to support the belief." [31]

Rejecting belief in an all-powerful Creator as faith-based "religion," while embracing a luck-of-the-draw invention of the human mind as "science"--epitomizes intellectual hypocrisy.

Regardless of the label, marketing a poison pill as a vitamin does nothing to enhance public health.

"The chances that life just occurred are about as unlikely as a typhoon blowing through a junkyard and constructing a Boeing 747." [32]

Life does not create itself from non-life---not millions of years in the past! Not today. Not ever.

Matter and energy don't result from "nothing" exploding.

When a computer crashes, intelligent information never uploads itself. With no place else to go, life resulting from a destructive cosmic explosion followed by an accident in Darwin's "warm little pond," retreats to the murky shadows of deep time's non-explanation.

Evolution fallacy is consistent only in its incoherent irrationality.

Thomas A. Edison recognized that "...this world is ruled by infinite intelligence. Everything that surrounds us – everything that exists – proves that there are infinite laws behind it. There can be no denying this fact. It is mathematical in its precision." [33]

Werner von Braun added a similar take:

"I find as difficult to understand a scientist who does not acknowledge the existence of a superior rationality behind the existence of the universe as it is to comprehend a theologian who would deny the advance of science." [34]

IV

Superstitious Nonsense

Spontaneous Generation?

"To get a cell by chance would require at least one hundred functional proteins to appear simultaneously in one place. That is one hundred simultaneous events each of an independent probability, which could hardly be more than 10^{-20} giving a maximum combined probability of 10^{-2000}." [1]

Michael Denton

Historic Old Colorado, John Herbert Johns, unpublished portfolio

"Carl Sagan estimated that the chance of life evolving on any single planet like, the Earth, is one chance in $1x10^{2,000,000,000}$...It would take 6,000 books of 300 pages each just to write the number." [2]

"A number this large is so infinitely beyond one followed by 50 zeroes (Borel's upper limit for such an event to occur)...There is, then according to Borel's law of probability, absolutely no chance that life could have 'evolved spontaneously' on the Earth." [2]

Charles Robert Darwin lacked the faintest clue of scientific evidence explaining the origin of first life here. Still, he didn't hesitate launching a series of dubious explanations built on the assumptive origin of that first living cell creating itself from inorganic chemicals.

"All the higher mammals are, "he wrote, "probably derived from an ancient marsupial, and this through a long line of diversified forms, either from some reptile-like or some amphibian-like creature, and this again from some fish-like animal." [3]

More than 150 years after he went public with his speculations, the question of how the first living cell managed to emerge accidentally from non-living matter continues to confound. Even as we are blessed with research technology unavailable to Darwin, flawed logic plagues evolution theory.

Darwin had never heard of DNA, or even a cell's nucleus, so he waved his wand imagining that the magic of self-creation took place magically in that "warm little pond." [4]

And evolution's theoretical beginning is not faith based?

The "pond" scenario seems less likely than finding an iceberg floating in a desert mirage.

Spontaneous generation of a living cell from non-living inorganic matter overwhelms imaginations. Just how could a cell, without evidence of prior ancestry, arrive from nowhere, pre-loaded with genetic information?

Richard Hutton, Executive Producer of the controversial PBS TV series, "Evolution," was asked, "What are some of the larger questions still unanswered by evolutionary theory?"

He replied: "The origin of life. There is no consensus at all here---lots of theories, little science. That's one of the reasons we didn't cover it in the series.

"The evidence wasn't very good." [5]

Maybe "worthless" rather than "not very good?"

Lacking the first clue as to the what, how, when, where and why, of first life's origin, Darwin conceded, "Science as yet throws no light on the far higher problem of the essence or origin of life." [6]

The best Darwin could do to explain this "far higher problem" was to side-step the non-explanation's dubious adequacy. Undeterred, he chose to rock the 19th century world by releasing his 1859 edition of *Origin of Species*.

Claiming all life began with that "cell from nowhere," he imagined the unexplained event was followed by a fanciful series of incremental changes over large chunks of deep time.

The chain of do-it-yourself "biological transit stops" supposedly produced every known plant and animal species, eventually replacing all ancestors radically new and different body plans.

Kids understand that two atoms of hydrogen joined to a single atom of oxygen build a water molecule. Yet minds swearing allegiance to the spontaneous generation fallacy are unable to conjure up the recipe mix for the quite hypothetical and life-creating, "prebiotic soup."

Darwin allied his faith with philosophers like Aristotle, who preached spontaneous generation as some pagan brand of superstitious gospel. Then along came French scientist Louis Pasteur with scientific demonstrations that shattered the spontaneous generation myth.

"The origin of life by chance in a primeval soup is impossible in probability in the same way that a perpetual motion machine is impossible in probability." [7]

Today, the formula for first life continues to baffle devout evolutionists, who can't begin to replicate the simplest cell and are at a loss to account for the source of genetic information packed into that first cell's DNA.

Lacking elsewhere to turn, evolution's more dedicated shills tout outer space as a possible ultimate source of life on earth. Even if plausible, a pair of problems looms: How could a life form survive the hazards of space travel and how did it manage its own life's accidental kick-start to begin with?

Evolutionism's recipe for life's "Formula One" evades hot pursuit. By default, it is left to Mother Nature's whims and Pappy Time's antiquity to parent first life.

With even Darwin questioning "chance" as a serious factor in "the world as we know it," why not confront the ultimate challenge to human intelligence and create life in a lab from non-living, chemical compounds?

If life from unintelligent non-life could result theoretically from an accidental whim of nature, then why couldn't human intelligence be recruited to design and create a living cell by duplicating the secret of first life's launching pad?

The challenge beckoned audacious minds.

Without the magnifying power of electronic microscopes, 19th century scientists dismissed a living cell as nothing more than a protoplasm blob.

Then along came Gregor Mendel and his garden of colorful blossoms, introducing science to the world of genetics--a world that would soon stagger conventional thought.

That presumed simple "blob" opened eyes to unimagined complexity.

What leaped into view through the lens of the electron microscope was a throbbing piece of molecular machinery, complete with a nucleus packed with genetic information--a variety of proteins, all wrapped in a membrane and surrounded by a cell wall.

"The tiniest bacteria cells are incredibly small, weighing less than 10^{-12} gms, each is in effect a veritable microminiaturized factory containing thousands of exquisitely designed pieces of intricate molecular machinery, made up altogether of one hundred thousand million atoms, far more complicated than any machine built in the non-living world...

"Nor is there the slightest empirical hint of an evolutionary sequence among all the incredibly diverse cells on earth...

"The complexity of the simplest known type of cell is so great that it is impossible to accept that such an object could have been thrown together suddenly by some kind of freakish, vastly improbable event. Such an occurrence would be indistinguishable from a miracle." [8]

The single-cell *Mycoplasma,* a microorganism without a cell wall, the "simplest known self-reproducing life form," carries 482, life-directing genes.[9]

The chance of spontaneous generation producing the complete formula of molecules, amino acids, and proteins essential for a cell only one-tenth the size of *Mycoplasm hominis H. 39*, is less than one in $10^{340,000,000}$. [10]

So the stage was set for a scientific challenge pitting nature's "chance" versus mortal intelligent "design." Could something similar to *Mycoplasma* be reproduced in a laboratory under the auspices of human minds?

Stanley Miller and Harold Urey stepped to the plate in 1952, intent on answering the challenge---only months before the DNA double helix string of information housed in living cells grabbed international headlines. This innovative duo shaped their experiment following a theoretical trail blazed by Russia's Alexander I. Oparin's and Britain's J.H.S. Haldane's attempt to rescue spontaneous generation ideology from history's dust bin of trashed ideas.

Adding intelligence to the formula struck at the heart of an equation devised to authenticate impossible accident. Even if successful, the intervention of human thought sabotaged the credibility of a process intended to confirm the non-intelligent origin of life.

Amino acids, the building blocks linking cell proteins, have been synthesized in laboratory environments. But the creation of a full complement of proteins essential to life, from laboratory-built amino acids, proved futile as attempting to break the sonic barrier riding a broomstick.

There is more: Amino acids can't form in the presence of oxygen.

Bypassing this hurdle, the Oparin-Haldane hypothesis theorized earth's original atmosphere maybe consisted of carbon dioxide, carbon monoxide, ammonia, methane, hydrogen and water vapor---but without oxygen. Oparin-Haldane's scientific stretch supposed inorganic "...chemicals combined to form organic compounds, such as amino acids, which in turn combined to form large, complex molecules, such as proteins, which aggregated to form an interconnecting network and a cell wall." [11]

Applying this formula as a basis for manufacturing life in a laboratory, the Miller/Urey team attempted to create a reducing atmosphere by circulating a high-energy spark through methane, ammonia, and hydrogen gases and a circulating hot water vapor.

The process produced "a small mass of black tar" along with "a condensed red liquid" containing some amino acids. Still, the nagging problem persists: The experiment "...only works as long as oxygen is absent and certain critical ratios of hydrogen and carbon dioxide are maintained..." [12]

Subsequent experiments using ultraviolet radiation produced "nineteen of the twenty biological amino acids and five nucleic acid bases of DNA and RNA." [13]

Miller/Urey required an elaborate trail of happenstance, with each step unaccountably fostered in the prebiotic soup. They counted on energy from lightning, earthquakes, volcanoes, and the sun's rays to trigger chemical reactions with atmospheric gases such as methane, ammonia, hydrogen, ethane and water vapors, conveniently converting them to amino acids, fatty acids and sugars.

Relying on the luck of the draw, they thought that these compounds could theoretically link up to form larger protein and DNA molecules, ultimately becoming "the first true cell" capable of "metabolism, genetic coding, and the ability to reproduce" when wrapped with a membrane. [14]

One public school text, relying on "could have" and a series of "may haves," misled students with a string of imaginative speculations.

"...Primitive earth may have had an atmosphere largely of hydrogen which was later lost to space. A secondary atmosphere may have included ammonia, methane, water, and hydrogen sulfide...Ultraviolet light from the sun, electrical storms, and decay of radioactive elements may have provided the energy to combine these molecules as sugars and amino acids.

"Amino acids could have combined to form proteins..." [15]

"May have" and "could have" don't disguise speculation.

Could lightning, heat from volcanoes and the sun's ultraviolet rays have actually "...affected gases in the primitive earth's atmosphere and changed them into more complicated organic compounds..." such as fatty acids, amino acids, sugars, and nucleotides? And then "accumulated in the ocean and then linked up with each other to form very complex molecules..." such as lipids, peptides, carbohydrates, polynucleotides and eventually combined to form "complex proteins?" [16]

"Urey and Miller assumed that methane was plentiful in the early earth's conditions. If this is true, the sun's ultraviolet light would have caused hydrocarbons to form and absorb in the clay at the bottom of the ocean. The deposits from Precambrian periods should then contain significant hydrocarbons or remains of carbons, as well as some nitrogen containing compounds.

"None of these are present in these deposits." [17]

Without a reducing atmosphere the Oparin-Haldane hypothesis for first life had no chance to begin. But without oxygen, life on Planet Earth could not exist.

"...Oxygen is necessary to protect proteins and DNA from the sun...Living organisms [bacteria] and organic molecules [amino acids, proteins and DNA] need the protection from ultraviolet radiation provided by an ozone screen [which is derived from oxygen]. [18]

"...If even trace amounts of molecular oxygen were present, organic molecules could not be formed at all." [19]

"Since living organisms [bacteria] and organic molecules [amino acids, proteins] need the protection from ultraviolet radiation provided by an ozone layer [which is derived from oxygen] yet the presence of oxygen, [in the atmosphere] prevents the development of such living systems and biological molecules [amino acids], this would constitute a catch-22 in the model." [20]

Oxygen is the critical component of today's atmosphere. Oxides in the rocks suggest oxygen was present also in ancient atmospheres.

"Iron oxide minerals have been found in Greenland, dating to 3.9 billions years ago. The presence of oxides suggests that oxygen was present at the time." [21]

Overeager celebrants initially interpreted the result of Miller/Urey's experiment as a break-through in creating virtual life from non-life in a test tube. Realists recognized much less. Brilliant human minds hit the wall---over their heads in a realm reserved for a "Superior Rationality."

The Miller-Urey experiment faced "...withering criticism from chemists for ignoring the role of competing and destructive cross-reactions with chemical ions that would be expected in any hypothetical ocean or pond. These reactions would have tied up or terminated any growing polymer-chain." [22]

The brilliant scientist team understood innovative human genius had failed to create life. Nor did they demonstrate the simplest living cell could have originated spontaneously from inorganic matter. To this day, finite minds, capable of creating computers, have never successfully designed and built a living cell from non-living, chemical matter.

Early in 2011, a team of Japanese scientists announced a five-year plan to access cloning technology in an effort to introduce the long-extinct mammoth to the twentieth century. The plan calls for replacing the nuclei from an elephant's egg cell with DNA extracted from mammoth tissue preserved in a Russian lab, and then placing the redesigned egg in an elephant's uterus.

Whether or not successful, cloning, like gene splicing, is not the creation of new life from non-living matter but is the result of human intelligence designing life forms by mixing precisely selected DNA with pre-existing living tissues.

Redundant propaganda touting unproven assumptions as scientific "fact" camouflage congenital defects plaguing evolution dogma.

Assertions built on assumptions neither manufacture nor equate fact. Superstitious nonsense blossoms when the assumption virus invades reasoning. Clinging to assumptions as fact is a faith exercise. Science, like religion, can be vulnerable to distortion by assumption.

Free society media champions free speech in pursuit of truth. Still, subtle enticements beckoning from evolution's "working hypothesis," can turn scramble fact with fiction. *USA Today* fell prey to the trap. In its August 9, 2005 edition, it exalted evolution, taking a swipe at "Intelligent Design," a nemesis to the troubled chance hypothesis.

"It [evolution] is the cornerstone of modern biology. Though there are various 'missing links' in the evolutionary chain, it has never been refuted on a scientific basis." [23]

It's strange irony that a team of intelligent media minds composed phrases, designed a layout, printed and distributed thousands of copies while believing human brains created themselves, by random accident, from some undetermined, unintelligent source--without design or designer.

What precisely did Darwin "postulate" that "scientists have confirmed?"

Brushing ponderous phrases aside, if evolution's postulates were a publicly traded stock, its pending market collapse should be imminent.

British scientist Gerald A. Kerkut acknowledged evolution is riddled with unproven assumptions that "...by their nature are not capable of experimental verification." Topping the list of seven is the assumption "that non-living things gave rise to living material, i.e. spontaneous generation occurred." [24]

Kerkut's six other eye-popping assumptions do nothing to fortify evolution's credibility as "confirmed" science.

"The second assumption is that spontaneous generation occurred only once...The third assumption is that viruses, bacteria, plants and animals are all interrelated. The fourth assumption is that the Protozoa gave rise to the Metazoa.

"The fifth assumption is that the various invertebrate phyla are interrelated. The sixth assumption is that the invertebrates gave rise to the vertebrates. The seventh assumption is that within the vertebrates the fish gave rise to the amphibia, the amphibia to the reptiles, and the reptiles to the birds and mammals..." [24]

Kerkut's roster of assumptions threaten the fabric of evolution's grand scheme, nagging at the fringes of Darwinian thought. His candid assessment concludes, "The 'General Theory of Evolution' and the evidence that supports it is not sufficiently strong to allow us to consider it as anything more than a working hypothesis." [25]

Evolution's legacy teeters on this pile of assumptive sand. Imaginative assumptions are inadequate substitutes for verifiable evidence. Assumptions become presumptions providing thin cosmetic cover for conjecture. Until verified by evidence, assumptions melt like wax caressed by probing rays of the noonday sun.

Nothing of significance has changed since Kerkut's reality checks.

Post-1859 breakthroughs have yet to rescue Darwin's ideas from its congenital "flaws" and "holes" in order to justify christening chance hypothesis speculation as a "fundamental fact of biology."

British evolutionist, Sir Fred Hoyle concluded, "the notion of life---with its incredibly intricate genetic code---originating by chance in some sort of primordial organic soup, was...'nonsense of a high order.'" [26]

A class-five hurricane scrambles and devours houses, trees and telephone lines marking its fickle path of raging fury. No one has seen this fearsome force of nature, shape the jumble of scrap residue, strewn helter-skelter in its wake, design and create a fully functioning personal computer. The chance of a hurricane's force creating a computer seems less farfetched than imagining a cell spawned itself spontaneously in a warm water pond.

Sir Francis Crick confessed: "An honest man, armed with all the knowledge available to us now, could only state that in some sense, the origin of life appears at the moment to be almost a miracle, so many are the conditions which would have had to have been satisfied to get it going." [27]

Louis Pasteur demonstrated the vacuity of the belief that inorganic matter spontaneously generates life. Pasteur's findings cut the heart out of naturalism's taproot, confirming that life begets life and like begets like. He wrote: "...The impossibility of the appearance of life from non-living matter...To bring about spontaneous generation would be to create a germ. It would be creating life...God as author of life would then no longer be needed. Matter would replace Him." [28]

Hardly a year had passed after publication of *Origin of Species* before Darwin admitted to second thoughts.

Writing to his American friend and Harvard botanist Asa Gray, the aging Darwin dropped a verbal bombshell. After proposing a theory built on "chance," evolution's big gun, still stubbornly refusing to acknowledge "design" in nature, confessed to finding his thinking in a "hopeless muddle."

"I cannot think that the world, as we see it, is the result of chance, and yet I cannot look at each separate thing as the result of Design. I am conscious that I am in an utterly hopeless muddle." [29]

Darwin's pause for second thoughts was justified.

Something was strangely out-of-kilter with the supposition that a living cell was nothing more than a blob of protoplasm, capable of creating itself, by chance, from non-living chemicals.

More than a century later, molecular biologist Michael Denton set the record straight, dismissing chance in the cell formation process. Denton characterizes spontaneous generation of a living cell as both "freakish" and "vastly improbable."

More than the simplistic blob perceived in Darwin's day, Denton saw a mechanism more complex than any man-made machine. He reasoned it would be "indistinguishable from a miracle" if a cell had been "thrown together suddenly by some kind of freakish…event." [30]

"It is the sheer universality of perfection, the fact that everywhere we look, to whatever depth we look, we find an elegance and ingenuity of an absolutely transcending quality, which so mitigates against the idea of chance." [31]

Aware of the primitive nature of scientific technology in the mid-nineteenth century, Denton cut nineteenth century scientists some slack.

"There is little doubt that if this molecular evidence had been available one century ago it would have been seized upon with devastating effect by the opponents of evolution theory…and the idea of organic evolution might never have been accepted." [32]

Mathematically, the odds against life by "chance" equate impossible.

Regardless, media patrons of the chance hypothesis tend to flak, hype, and shill evolution's obsolete myths in a blizzard of clichés, confusing the public by awarding sham "science" an undeserved place of honor in the pantheon of academic respectability.

Mark Twain's tongue-in-cheek humor still resonates:

"There is something fascinating about science. One gets such wholesale returns of conjecture out of such a trifling investment of fact." [33]

V

Paging Sherlock Holmes

Fossil Mania

"To take a line of fossils and claim that they represent a lineage
is not scientific hypothesis that can be tested, but an assertion
that carries the same validity as a bedtime story—
amusing, perhaps even instructive, but not scientific." [1]
Henry Gee

Historic Old Colorado, John Herbert Johns unpublished portfolio

Darwin predicted “finally-graduated organic chain[s]” of transitional life forms must exist in fossil cemeteries if his “theory be true.”

Charles Darwin recognized his theory teetered in serious jeopardy unless it were confirmed by an “inconceivably great” quantity of “transitional links” yet to be discovered in the fossil record.

“If my theory be true, numberless intermediate varieties, linking closely together all the species of the same group, must assuredly have existed” [2]

“The number of intermediate and transitional links between all living and extinct species must have been inconceivably great.” [3]

Many million fossil finds later, the cupboard of “transitional links” remains embarrassingly bare. Using Darwin’s own rules, the game should have been over.

Diehards haven’t given up; the goal posts keep getting moved.

With hardly a hint of persuasive evidence, Charles Darwin spun a tale of a “finely graduated organic chain” [4] of life evolving gradually from that first living spark to a plethora of diversified complexity in a grandly conjectured “tree of life.” Choice words highlighted unproven assertion.

Optimistic predictions have not fared well, exposed to the shovels and microscopes investigating post-Cambrian evidence. In a game of paleontological hide-and-seek, undiscovered fossil fragments of lives that never existed have yet to be found. Not apparent in living systems either, evolution’s last best hope for the spectacle of an “organic chain” is anything more than wishful thinking.

Intensive quests have turned up little but frustration’s heat. Undeterred, Darwin accessed his fertile imagination, conjuring up visions of never-seen, fictional critters intended to illustrate his radical idea. Pushing credibility to the limit, he introduced a single sex link to the mythical chain. "Some extremely remote progenitor of the whole vertebrate kingdom appears to have been hermaphrodite or androgynous." [5]

Ultimately, the organic chain of quaint imaginings gradually drifted past the "vertebrate...hermaphrodite" phase leading quite accidentally to a dubious depiction of human ancestors pinned with a "tail" and a head crowned with "pointed ears."

"Man is descended from a hairy quadruped, furnished with a tail and pointed ears, probably arboreal in its habits..." [6]

"Early progenitors of man were no doubt once covered with hair, both sexes having beards; their ears were pointed and capable of movement; and their bodies were provided with a tail..." [7]

This is not make-believe movie madness, but Darwin's serious overreach designed to demonstrate the potential of his idea if given enough time, all descending from that original cell, which theoretically débuted on earth at a time and place he knew not how, when or where.

Contrary to evolution theory, the fossil record exposes multi-thousands of distinctly different plant and animal life forms that appear abruptly, without evidence of prior ancestry.

So what was the legacy of Darwin's "imagination?"

His bold conjecture called for the constant shifting of unstable life forms, moving in a series of blind transitions to unpredictable "biologic transit stops," climbing a non-existent taxonomic tree intended to serve both plants and animals as home base.

That first life supposedly progressed gradually up the taxonomic ladder through a yet-to-be-proved evolutionary sequence: "Fish-like," "Aquatic animal," "Amphibian-like creature," "Reptile-like," "Higher mammals," "Old World division of the Simiadae;" with "Man" emerging victorious atop of the heap.

Reality destroys this fallacious assumption. Reliable evidence of an organic chain of ancestral life forms, transiting one kind-to-a new and different kind, doesn't exist.

Genetic adjustments, within a life-form kind, can be real but the genetic changes work down rather than up the taxonomic ladder. At least 7,640 fully-formed, mostly marine animal species, appeared worldwide, in a simultaneous "sudden leap," saturating the Cambrian. [8] The "almost abrupt appearance of the major animal groups," representing as many as 50 phyla during the Cambrian Period, ranks as "...one of the most difficult problems in evolutionary paleontology..." [9]

Cambrian fossils cripple Darwin's primary premise. Unrelated phyla appear abruptly, across-the-board, rather than evolving gradually, starting with a single cell. [10]

Multi-celled animal embryos, no bigger than a grain of sand, have been discovered in China, preserved in calcium phosphate and dated at the edge of the Cambrian/Precambrian time frame—marking the prolific explosion of organic life during which "virtually all the major animal body plans seen on Earth today blossomed in a sudden riotous evolutionary springtime." [11]

The Cambrian explosion of multi-celled organisms covers the waterfront as to invertebrate animal phyla, both living and extinct. Arthropods, mollusks, and echinoderms—there they are, without a fossil clue as to prior "numerous, successive, slight modifications."

"Most orders, classes, and phyla appear abruptly, and commonly have already acquired all the characters that distinguish them." [12] "For all of the animal phyla to appear in one single, short burst of diversification is not an obviously predicable outcome of evolution." [13] "...Radically new kinds of organisms appear for the first time in the fossil record already fully evolved, with most of their characteristic features present." [14]

Henry Gee discounts evolution's supposed organic chain of ancestry as "...a completely human invention created after the fact, shaped to accord with human prejudices...Each fossil represents an isolated point, with no knowable connection to any other given fossil, and all float around in an overwhelming sea of gaps." [15]

Far from supporting descent from a chain of organic intermediates, the Cambrian and Pre-Cambrian fossil record offers a virtual "paleontological desert," barren of readily identifiable fossil ancestors.

Irreducibly complex animal and plant life forms descending in gradual increments from some original single-cell first life doesn't match Cambrian fossil reality. Choice rhetoric highlights unproven assumption but chunks of deep time can't account for leaps over vast chasms of biological diversity that never happened.

Cambrian fossils suggest fully formed life existed from the start, without evidence of prior fossil ancestry, and that many descendant species carry remarkable resemblance to theoretically extinct ancient ancestors.

Once the academic race was on for fame and fortune
in quest of ancient-bone evidence corroborating evolution,
a plethora of suspicious findings surfaced.

Paleontologists Edward Drinker Cope and Othniel Charles Marsh squandered personal reputations competing for fame and fortune in a "Bone Wars" vendetta, late in the 19th century. Their efforts harvested tons of dinosaur remains, strewn in fossil cemeteries across the high prairies of Colorado and Wyoming. Bitter rivalry raged between the two adversaries from 1877 to 1892.

No question, multi-ton finds of gigantic dinosaur bones confirmed extinction of a once monstrous species. But the petty dispute between two adversaries tainted professional careers and depleted personal fortunes.

Several-mile stretches of massive fossil cemeteries indicated where lush forests once stood. Marine fossils in the mix, at elevations as high as 7,881 feet in the general area of Morrison, Colorado, left evidence of what must have been rapid burial in Dakota sandstone. Sudden hydraulic action, without precedent in the modern world, prevented decay.

Cope and Marsh were not alone in straddling enormous dinosaur spines in quest of professional distinction and financial reward.

Paleontologist Hermann von Meyer led another pack of resourceful entrepreneurs, looking to cash in with the remains of *Archaeopteryx*, a pigeon-sized imprint of bones unearthed in Germany's Solnhoffen Quarry. Earlier, French paleontologist Geoffrey Saint-Hilaire, floated the idea that birds possibly evolved from reptiles. The dinosaur-to-bird concept ignored radical dissimilarities between the two animal types (eg. warm-blooded birds v. most reptiles are cold-blooded). Darwin loyalists, eager to bridge the fossil record's yawning gaps, seized upon the *Archaeopteryx* remains, etched in limestone slabs, as evidence of reptile-to-bird transition.

Skeptics, then and now, question the connection. Andreas Wagner, discoverer of a dinosaur fossil he named *Compsognathus*, dismissed the claim, warning: "Darwin and his adherents will probably employ the new discovery as an exceedingly welcome occurrence for the justification of their strange views upon the transformation of the animals. But they will be wrong." [16]

Alan Feduccia scoffed at the dinosaur-to-bird scenario.

"The theory that birds are the equivalent of living dinosaurs and that dinosaurs were feathered is so full of holes that the creationists have jumped all over it, using the evolutionary nonsense of 'dinosaurian science' as evidence against the theory of evolution... To say dinosaurs were the ancestors of the modem birds we see flying around outside today because we would like them to be is a big mistake." [17]

So was *Archaeopteryx* just another extinct bird species? Or even a pigeon-sized dinosaur? Since a *Pterosaur*, with a10-inch wingspan, was also discovered nearby in Sondheim, Bavarian limestone, could *Archaeopteryx* have been a relative?

No sooner had the twenty-first century dawned than the bird-to-dinosaur scenario took a hit from another counterfeit fossil. This time, the victim was much more than a bit player in the drama.

Photos of fossil remnants trumpeted as "...a missing link between terrestrial dinosaurs and birds that could actually fly...a true missing link in the complex chain that connects dinosaurs to birds" adorned full-color pages of the November, 1999 issue of the prestigious *National Geographic*. [18]

Big bucks backed the journal's investment in "authentic" antiquity. The endorsement came loaded with monumental consequences given the journal's multi-million circulation and its proud tradition of editorial excellence.

But despite the pretty pictures and the exuberant tone implying authenticity, by January of the new millennium the speculative dreams of the sponsor came crashing to the earth, as embarrassingly flightless as lifeless. The costly find, imported from China's Liaoning Province, proved to be nothing more than a carefully concocted swindle: Some enterprising huckster hitched the tail of a long-deceased dinosaur onto the fossil remains of an equally unsuspecting bird. The only thing the dino/bird fossil combo achieved was the egg it smeared on the faces of over-eager evolution enthusiasts.

Lemmings march mindlessly off cliffs in a mysterious dance of death. Remarkably, the marvelously more sophisticated human mind can also be vulnerable to lock-step brands of cultural and academic coercion.

A French game show instructed contestants to jolt fellow contestants with up to 450 volts of electricity when they heard wrong answers. Despite the anguished cries of the "victims," four of five participants reacted to noisy shouts from the audience, pushing them to inflict unconscionable pain on fellow humans. The results delivered the "shock!"

Unaware the event was a sham, without surging voltage, and that writhing "victims" were actors selected for a scientific test, all but 16 of the 80 contestants followed instructions mechanically, irrespective of the perceived "hurt" they inflicted.

Not exactly a vote-of-confidence as to freethinking independence and gracious sensitivity of the human race when peer pressure pushes the buttons of psychological coercion.

Flawed dogmatic traditions impact politics, history, religion and science.

© Mike Norton

Authoritarian, coerced conformity can infect classrooms. When respected professors portray evolution as "fact," survival instinct kicks in motivating students to shrug and bite the hook. The most astute human minds are susceptible to rote tradition and the intense, repetitious touting of a party line. Totalitarian political regimes exploit this shortfall in human psyches.

Does this partially explain why loyalists continue to pledge allegiance to Darwin's flawed dream after the natural limits to biological change have been apparent for more than a century?

Darwin predicted his theory would "absolutely break down... if any complex organ...could not possibly have been formed by numerous, successive, slight modification." [19]

"Nowhere was Darwin able to point to one bona fide case of natural selection having actually generated evolutionary change in nature." [20]

If Darwin's reasoning is to be taken seriously and ancient-time evolution is real, with life discounted as nothing more than a transitory accident, every fossil bone chip should represent a link in a continuous chain of organic life.

This is not the case. Evidence of radical transformation in life formats, by drawn-out whims of "gradualism," ranges from the suspiciously fragile to the non-existent.

Fast forward to century twenty-one and millions of fossil discoveries later, the "most obvious and serious objection...against the theory," lingers in high-definition, academic shadows, an unresolved nemesis.

With more than a couple billion bits and pieces of fossil bone fragments or imprints in the earth's crust having already surfaced, and an estimated 200,000,000 fossils displayed in world museums, it should be a slam dunk to match Darwin's dream with science reality---if Darwin's theory really were true.

Fossil graveyards should overflow with billions of intermediates, transiting to radically new and diversified organic formats---if the theory really were true.

Late in the twentieth century, articulate evolutionist, Harvard's Stephen Jay Gould, recognized the shortfall and spoke with academic candor.

"The absence of fossil evidence for intermediary stages between major transitions in organic design, indeed our inability, even in our imagination, to construct functional intermediates in many cases, has been a persistent and nagging problem for gradualistic accounts of evolution." [21]

A year later, evolutionist Colin Patterson confirmed reality.

"Gould and the American Museum people are hard to contradict when they say there are no transitional forms.... I will it lay it on the line—there is not one such fossil for which one could make a watertight argument." [22]

Instead, with the weight of evidence leaning heavily in the direction of nothingness, "break down" time knocks at the chance hypothesis door. Those billions of partially built, transiting plant and animals, haven't been found.

There is nothing in the living world that demonstrates evolution-in-action. Discontinuity reigns. Gaps dominat. So where does that leave neo-Darwinism's molecule-to-man nonsense?

Confined to a whiff of time and a sliver of space, humans reach for evidence correlating our genesis with today and an infinity of tomorrows.

Evolution isn't happening now as a newsworthy current event; it hasn't happened in the recent past; it never happened. The fictitious scenario features a counterfeit "science" that corrupts minds with a contagious intellectual virus.

More than an academic charade, belief in life's origin comes with consequences: Purpose, quality of life and destiny impact all humans. Life is God's gift, now and forever. To deliberately brush aside and ignore this miracle demonstrates the incomprehensible arrogance of ignorance.

"...Evolution is the central most disorganizing, anti-intellectual anti-science principle that biologists have ever been dictatorially forced to learn to understand the world...it stands as the greatest scandal in science of the last 140 years." [23]

Evolution devotees may ignore the missing evidence, insisting the obvious gaps are exceptions to the "rule." Doubters are more likely to perceive the glaring exceptions tend to "eat up the rule."

As to that elusive *"organic chain"* of successor intermediates, the idea "carries the same validity as a bedtime story—amusing, perhaps even instructive, but not scientific." [24]

Where is the mythical Sherlock Holmes when he's needed to locate those still missing bits of ancient bones? Even the legendary Holmes couldn't pull mythical fossils out of sedimentary rock.

In the phrase made famous by Sherlock, it's "elementary, my dear Watson."

VI

"...According To Its Kind" [1]

Stasis

"We may safely infer that not one living species will transmit its unaltered likeness to a distant futurity ...." [2]

Charles Darwin

© Clara

This giant fossilized nautilus shows a striking resemblance to its twenty-first century descendants.

Darwin predicted "not one living species will transmit its unaltered likeness to a distant futurity." The fossil record contradicts his prediction.

The Pre-Cambrian landscape presents a "paleontological desert." Rather than a reliable geologic column brimming with obvious transitional, stasis intrudes. Today's world is loaded with look-alike descendants of Cambrian animal and plant species, relatively unchanged from their fossil ancestors, undercutting an already tenuous theory predicated on an organic chain of persistent transitions.

"Almost abrupt appearance of the major animal groups," from as many as 50 phyla, ranks as "one of the most difficult problems in evolutionary paleontology." [3]

This is "not an obviously predicable outcome of evolution." [4]

Discontinuity and stasis trump gradualist evolution.

The abrupt and relatively simultaneous appearance of several thousand different plant and animal species, without evidence of transition ancestry prior to the Cambrian, dealt a severe blow to the heart of evolution theory.

In addition to the once-considered extinct coelacanth, fossil cemeteries include mirror images of modern crustaceans, millipedes, horseshoe crabs, snails, bivalves, squid-like belemnolds, nautilus-like ammonolds and mollusks. Moving up the "geologic column" from the deepest sedimentary layers, we find ancestral duplicates of today's stromatolites, brittlestars, sea lilies, dragonflies, sycamore tree leaves, wasps, and sand dollars abound.

Multi-celled animal embryos, no bigger than a grain of sand, have been discovered in China, preserved in calcium phosphate and dated at the edge of the Cambrian/Precambrian time frame—marking the prolific explosion of organic life during which "virtually all the major animal body plans seen on Earth today blossomed in a sudden riotous evolutionary springtime."[5]

Henry Gee's scrutinizing eye views attempts to "take a line of fossils and claim that they represent a lineage" as a less than scientific "assertion that carries the same validity as a bedtime story." [6]

"All fossil roundworms found to date are indistinguishable from modern forms." Fossils conventionally dated "more than 3 billion years old are virtually idntical to organisms, called cyanobacteria, living today." [7]

Shixue Hu of China's Chengdu Geological Center reports discovering a cemetery of 20,000 fossils, many fully intact, buried fifty feet deep in a Luoping mountain in southwestern China. Duly awarded a 250 million year conventional age, the treasure trove of antiquity includes stunning evidence at odds with Darwin's prediction that "not one living species will transmit its unaltered likeness to a distant futurity." [8]

There they were, 250 million conventional years after the fact: clams, oysters, snails, virtual replicas of life forms that inhabit today's earth.

Stasis prevails.

The find's stunner was the "soft tissues" found on some of the fossils! [9]

"Soft tissues" that escaped decay for 250 million years? That's not supposed to happen.

No doubt about it, many plant and animal species have suffered extinction. Equally certain, a cast of thousands thrives profusely, perpetuating resemblance to ancient ancestors. Evidence preserved in stone contradicts Darwin's guess that "not one living species" would escape to the future. The fossil record didn't oblige his expectations.

Instead of a reliable geologic column brimming with the intermediates mandated by the chance hypothesis, stasis intrudes. The fossil record bulges with evidence of species transmitting their "unaltered likeness to distant futurity," while persuasive evidence of fossil chains linking molecule-to-man remains sparse if not invisible.

Darwin reasoned, given enough time, every parent species would be wiped out by superior descendants. Conjectured chunks of deep time made anything seemingly possible--even exaggerated leaps over vast chasms of biological diversity.

© Megan Gayle

This horeshoe crab is a dead ringer look-alike for his ancient, fossilized ancestor.

To build his thesis on the theme that"the production of new forms has caused the extinction of about the same number of old forms," [10] Darwin had to turn a blind eye to overwhelming discontinuity in the fossil record. Contrary to his fantasy, millions of fossils shout stasis!.

After 425 million years of conventional time, the Ostracode, remains unchanged. [11] Ancestral alligators, oysters, sea urchins, horseshoe crabs, bowfins, Australian lungfish, sturgeons, crinoids, bats, arrow worms, opossums, star-fish, corals, and the platypus weave a consistent pattern of resemblance with descendants.

Out-of-kilter with Darwin's idea of "new forms" from the old, an allegedly 360 million-year-old fossil shrimp, discovered in Oklahoma, appears to be a duplicate of shrimp living on today's seafloors. [12]

Devonian era fossils, virtually identical to animal species observed in the twenty-first century world, not only confirm stasis but also call into question evolution's conventional claim that amphibians descended from fish ancestors. [13]

Evolutionist Henry Gee's ranked the horseshoe crab as yet another species that not only dodged extinction's bullet but also evaded marching in any parade of transitional forms. Gee described the crab as "an animal that has not changed its basic form for hundreds of millions of years." [14]

Steven Austin, devout creationist with impeccable science credentials, confirms that "Clams have always been clams; brachiopods have always been brachiopods; fish have always been fish." [15]

Even the lowly cockroach survives unscathed from mega-time's ravages.

The Western Pacific's nautilus and its long-fossilized ancestor appear to be twins. "In every way they are virtually identical to the living chambered nautilus. The creature that swims in our oceans today is the same one that was swimming 100 million years ago." [16]

Fossil insect species identified in ancient Scandinavian amber bear a striking resemblance to today's descendants.

"Lungfish almost identical to those of modern Africa are found as fossils in the rocks of the Devonian era…alongside fossils of the earliest amphibians and the very fish groups from which the amphibia supposedly arose." [17]

Greenling damselflies, tiny insects with 22mm wingspans, supposedly went extinct 250 million years ago, but there they are, still flying in Australia. [18]

While Darwin imagined extinctions, 1840's era excavators for England's Great Western Railway stumbled across a fossil cemetery "where thousands of Jurassic fossils with preserved soft tissues were found" and then lost until rediscovered by a team led by Phil Wilby. "Literally millions of these animals were dying out or being killed in this precise area and we don't know what that is." [19]

Not only did the ancient squid resemble its modern counterpart, but also its inch-long ink sac delivered the formula for the ink used by an artist to sketch its design and to sign its scientific name.

Wilby marveled, writing this: "It is difficult to imagine how you can have something as soft and sloppy as an ink sac fossilized in three dimension, still black, and inside a rock that is 150 million years old. The structure is similar to ink from a modern squid so we can write with it." [19]

A fossil spider with a one-inch body and legs stretching out 2.5 inches has been discovered in China's Inner Mongolia. Reputed to be 165 million years old by conventional standards, the arachnoid dubbed *Nephila jurassica*, is a match for today's tropical "golden orb weavers." [20]

Where is the evolution? If the alleged daring were accurate, this little guy with fuzzy fibers on its legs, argues for 165 million years of stasis rather than the faintest hint of evolution!

The paucity of fossil evidence corroborating his theory troubled Darwin. He recognized fossils found during his lifetime revealed few, if any "finely-graduated organic chain." He set the bar high, putting all his cards on the table with one giant leap of faith. "If my theory be true, numberless intermediate varieties, linking closely together all the species of the same group, must assuredly have existed" [21]

Frustrated, Darwin bemoaned the glaring gaps between theory and fact. "Geological research…does not yield the infinitely many fine gradations between past and present species required on the theory…Why do we not find beneath this system great piles of strata stored with the remains of the progenitors of the Cambrian fossils?" [22]

The shortfall nagged at the naturalist's mind, leading him to admit he recognized the missing chain of transitional links was "the most obvious and serious objection which can be urged against the theory." [23]

Since the publication of *Origin*, millions of fossils have been rescued from rocky hiding places. But the missing organic chains of transitional links continue AWOL.

Beyond defining "the most obvious and serious objection which can be urged against the theory" he acknowledged that "if it could demonstrated that any complex organ existed, which could not possibly have been formed by numerous, successive, slight modification, my theory would absolutely break down." [24]

To validate evolution theory, all life forms would be transiting by gradual increments of "successive, slight modification," and the fossil record would be brimming with intermediate life forms.

Instead, the fossil cupboard overflows with extinctions and stasis pointing to discontinuity rather than the requisite "inconceivably great" number of "transitional links."

Henry Gee acknowledges the chasms of discontinuity. "It is impossible to know for certain whether one species is the ancestor of another." [25] A realist, he points out "...that adaptive scenarios are simply justifications for particular arrangements of fossils made after the fact, and which rely for their justification on authority rather than on testable hypotheses." [26]

Ernst Mayr agreed, sensing discontinuity. "All species are separated from each other by bridgeless gaps; intermediates between species are not observed." [27]

Unwilling to discard his lifetime toil, Darwin backtracked by suggesting nineteenth century fossil discoveries must be woefully incomplete. He counted on future discoveries of fossil evidence to salvage his theory. Fortified with a visionary's optimism, Darwin shrugged off the shortfall, ascribing the missing branches in his conjectured "tree of life" to the "extreme imperfection of the geological record." [28]

Discoveries since brought little comfort to his dream.

The naturalist ignored bleak reality. His "inconceivably great" number of "transitional links" haven't been discovered yet, after all this time. Early in century twenty-one, with more than two billion fossils exposed to the light of day, those elusive "transitional links" continue missing in action.

Regardless, no amount of wishful thinking can mask imagination's blank screen bias. "Surprising as it may seem, the only real evidence for the geological succession of life, as represented by the timetable, is found in the mind of the geologist and on the paper upon which the chart is drawn.

Nowhere in the earth is the complete succession of fossils found as they are portrayed in the chart." [29]

Scraping the bottom of the bone barrel, efforts in the quest for missing links continue fruitlessly. Rather than a roster of substantive evidence, evolution loyalists have had to settle for the rare and the dubious.

Using fossil bone scraps in an attempt to link birds to dinosaurs and whales to a land-based ancestor with a foot is a far cry from reliable evidence. The idea that some mammal with a foot evolved into whale status affronts the Genesis account that "...God created the great creatures of the sea and every living and moving thing with which the water teems, according to their kinds, and every winged bird according to its kind." [30]

Historic Old Colorado, John Herbert Johns, unpublished portfolio

Suggesting game birds transited from dinosaurs would have triggered roaring laughter from these huntsmen.

Darwin's imagination demeans and disrespects human life while fostering irreverent contempt for the Creator. Since the living world is hardly awash with definitive examples of evolution in action, it remains for the fossil record to deliver more than unsubstantiated conjecture.

Whether exposed to the shovels digging for clues in fossil fields, or to spotlight glare of microscopes rolling back the mysteries of molecular biology, Darwin's elaborate predictions have not fared well. Absent requisite transitional forms, Darwin's dream dangles in limbo, clinging to an unraveling thread. Continuous chains of organic life forms have yet to be forged from unrelated fragments of fossil bones.

If we borrow a quip from Mark Twain's humor, we could evolution's supposed mantle of authenticity is "greatly exaggerated."

Or in the phraseology of gritty Texas lingo, evolution's unproved assumptions qualify as "all hat and no cattle."

VII

Imagination Run Amok

Living Cell

"To get a cell by chance would require at least one hundred functional proteins to appear simultaneously in one place. That is one hundred simultaneous events each of an independent probability which could hardly be more than 10^{-20}."[1]

Michael Denton

This century old "horseless carriage" didn't create itself.
The car's intricate design doesn't compare to
the complexity of the simplest living cell.

The brightest, intelligent human minds have never been able to create a living cell from non-living matter.

Two years before the publication of *The Origin of Species*, Charles Darwin wrote to Asa Gray admitting that in the absence of supporting evidence one's "imagination must fill up the very wide blanks." [2] But even Darwin's innovative mind never came close to filling the gaps in his imaginings or grasping the complex rhythm of the simplest living cell.

Night skies showcase multiples of ten million bright lights set against a dark field of infinite space. Invisible to the human eye, the simplest cell imaginable boasts a format of "supreme technology and bewildering complexity" built from "about ten million atoms." [3]

Primitive wisdom envisioned cells as little more than infinitesimal pieces of cytoplasm, *"a* relatively disappointing spectacle appearing only as an ever-changing and apparently disordered pattern of blobs and particles." [4]

Lacking access to electricity or the magic of the electron microscope, and presumably without knowledge of the genetic information revealed in Mendel's garden, Darwin crafted an idea that has been patched-up and presented to today's world as the Modern Synthetic Theory of Evolution.

Unaware of the typical prokaryote cell's complexity with its nucleus, DNA, proteins and cell wall, Darwin imagined, incorrectly, that acquired physical traits could be transferred genetically to offspring by use or disuse.

Blind to his ignorance of cell composition, Darwin plunged ahead, publishing his imaginings in 1859, bypassing, as a "far higher problem," the incongruity of original life emerging spontaneously from non-living matter. This nagging lack of a rational naturalistic explanation for the origin of first life stretched beyond anything his mind could conjure-up.

Evolution's premise, built on life from non-life, is more than dilemma--scientifically it's catastrophic. Fantasy can't pass for intelligent design.

Darwin never claimed microbiologist credentials. But his instincts warned him of the unresolved design dilemma. Baffled by the lack of a viable scientific explanation for the sudden appearance of first life, he likely struggled with the implications of his hypothesis. What he admitted to be an "utterly hopeless muddle" remained unresolved throughout his lifetime.

Eventually, twentieth century molecular reality exposed the lifeless roots of his concocted "tree of life." Before his 1882 death, Darwin confessed, "...I am quite conscious that my speculations run beyond the bounds of true science." [3]

When he died, the murky "muddle" had failed to inject even the hint of first life into the sterile stick of a figurative tree. With reason, the naturalist ranked his grand scheme as "...a mere rag of an hypothesis with as many flaw[s] & holes as sound parts." [5]

Darwin's wanderings in a dark, intellectual wilderness offered latitude for gross error. Aware of the primitive nature of scientific technology in the mid-nineteenth century, microbiologist Denton cut Darwin some slack.

"There is little doubt that if this molecular evidence had been available one century ago it would have been seized upon with devastating effect by the opponents of evolution theory...and the idea of organic evolution might never have been accepted." [6]

Denton's vision of a living cell examined under the probing eye of an electron microscope opened vistas Darwin never imagined.

Throughout the first post-Darwin century, fossil bones anchored pre-history investigations, providing thin grist for Darwinian rhetoric. In one giant leap beyond nineteenth-century science's near-sighted focus on fossils, intricacies of the complex cell took center stage under electron microscope scrutiny.

The new science of life pushed the cell's design and function front and center. Molecular biologists had pulled back a curtain that had been concealing the previously unseen and unknown. Eyes of discovery marvel in wonder at a pulsating package of coordinated microscopic motors and machines driving life's core.

The cell emerged as a vibrant speck of organic life, more complex than any machinery yet designed by human intelligence. What once had been dismissed as bland bits of protoplasm jumped out as genetic treasure troves.

The insightful discoveries revealed by molecular biology heralded an unwelcome wake-up call for Darwin's dream. The chance hypothesis itself required serious reevaluation.

With the advent of molecular biology, the mysteries of life's origin shifted from obsession with bits and pieces of fossil bones to the microscopic mother lode of information --DNA's "Language of Life."

Thanks to the power of the microscope, only four years after the 1925 Scopes trial, H. G. Wells and Julian S. Huxley spotted "snakelike threads" writhing "slowly through the cell" which they called "mitochondria" while noting they could see nothing "inside the nucleus but a clear fluid." [7]

This imperfect glimpse of a cell's inner workings proved but a hint of things to come. Once the electron microscope debuted with magnification capability of a million times, researchers devoured an unfolding panorama of knowledge.

Short decades after *Origin of Species* aroused debate, molecular biology surfaced raising eyebrows and temperatures. Investigation of previously invisible life forms turned naturalistic interpretations of the cell's intricate life system upside down. The "unparalleled complexity" of a living cell fascinates sophisticated observers.

A cell consists of nitrogen, hydrogen, carbon and oxygen. It stores an encyclopedia of working knowledge and reproduces copies of itself. This miniscule mechanism of grandeur performs its assigned tasks while acting in concert with other cells in a system sustaining a unique life form.

The overwhelming discontinuity at nature's molecular level is highlighted dramatically in Michael Denton's *Evolution: A Theory in Crises!*

"To grasp the reality of life as it has been revealed by molecular biology, we must magnify a cell a thousand million times until it is 20 kilometers in diameter and resembles a giant airship large enough to cover a great city like London or New York.

"What we would then see would be an object of unparalleled complexity and adaptive design. On the surface of the cell we would see millions of openings, like the portholes of a vast spaceship, opening and closing to allow a continual stream of materials to flow in and out. If we were to enter one of these openings we would find ourselves in a world of supreme technology and bewildering complexity...

"The simplest of the functional components of the cell, the protein molecules, were astonishingly, complex pieces of molecular machinery, each one consisting of about 3,000 atoms...

"What we would be witnessing would be an object resembling an immense automated factory...larger than any city and carrying out almost as many unique functions as all the manufacturing activities of man on earth...a factory which would have one capacity not equaled in any of our own most advanced machines, for it would be capable of replicating its entire structure within a matter of a few hours."[8]

Beyond a pulsating pack of cytoplasm with a nucleus wrapped in a membrane, the cell contains machinery performing a multitude of actions every split second. Mitochondria, located within the cytoplasm, generate and store energy.

"Life is far more than chemicals, and building life immensely more complex than pasting carbon, hydrogen, oxygen, and nitrogen together in clever ways. Every coffin in the cemetery is filled with those same chemicals, but no one walks out in the morning."[9]

Not only is the transitional continuity predicted by evolution lacking in fossil fields, the molecular world of living cells shouts discontinuity.

"We now know not only of the existence of a break between the living and non-living world, but also that it represents the most dramatic and fundamental of all the discontinuities of nature…." [10]

"It is well established that the pattern of diversity at a molecular level conforms to a highly ordered hierarchic system. Each class at a molecular level is unique, isolated and unlinked by intermediates. Thus molecules, like fossils have failed to provide the elusive intermediates so long sought by evolutionary biology." [11]

No scientist has yet successfully synthesized life from inert non-life, nor explained the how, when and where first life originated by accident. "…The chemical reactions required to form proteins and DNA do not occur readily. In fact, these products haven't appeared in any simulation experiment to date." [12]

Biochemist Michael Behe, working with an assistant in the NIH lab, "…analyzed the supposed miracle of the first living cell coming into being by historical accident. 'What would you need?' they asked each other. 'You need a membrane, a power supply, and you need some genetic information. You need a replication system. And we kind of stopped and looked at each other.

"We said, 'Nah.'" [13]

A living cell's package of microscopic data reflects formulas as reliably predictable as any array of inorganic atoms depicted in the Periodic Table of the Elements.

Francis Crick and James Dewey Watson earned in 1953 their figurative spurs in the science hall of fame by uncovering DNA's delicate, double helix design. Identification of DNA's incredible design clouded further the "hopeless muddle" of evolution's assumptive imaginings.

Nineteenth century simple cell dogma vanished, existing now only in the lexicon of obsolete science and the minds of the uninformed. Darwin's day academics would have been astounded had they been introduced to the invisible intricacies of one-celled Prokaryote organisms, wrapped in a membrane encasing protein and a coded dose of DNA information.

Beyond a pulsating pack of cytoplasm with a nucleus wrapped in a membrane, a cell contains machinery performing a multitude of actions every second. It must absorb food, discard wastes, repair, replace, reproduce, and grow---all functioning pursuant to a built-in information code. [14]

At a minimum, a living cell requires a system of regulatory mechanisms; a constant supply of energy; an abundance of four nitrogenous bases; ribotide phosphates; twenty aminoacyl nucleotidates; deoxyribonucleic acid (DNA); DNA polymerase; and RNA polymerase. [15] Evolution's postulated radical transition of a life format to a new and entirely different plant or animal can't occur, unless it happens here.

Ribosomes craft chains of proteins from a smorgasbord of twenty amino acids. Proteins don't form naturally from chemicals. Never has "one single functional protein molecule" been discovered resulting from random chance processes. The twenty amino acid chains that make proteins can't order themselves.

Typically a protein consists of 500 amino acid chains. There are more than 30,000 distinct proteins. [16] It's a tall order to deliver 100,000 different proteins to the human body by random chance.

A smorgasbord of amino acid varieties provides the raw material from which proteins are built. "A protein may have many of each kind. A typical protein will have a few hundred amino acids...To make a protein that will do something useful, the cell has to get the right amino acids in the right order." [17]

Protein molecules constitute "the simplest of the functional components of the cell." Each of these molecular machines consists of "about three thousand atoms arranged in highly organized 3-D spatial conformation...

"The life of the cell depends on the integrated activities of... probably hundreds of thousands of different protein molecules...It would be a factory which would have one capacity not equaled in any of our own most advanced machines, for it would be capable of replicating its entire structure within a matter of a few hours." [18]

The key to proteins folding into specific, three-dimensional shapes is in the sequence and arrangement of amino acids. After assembling in correct sequence, a protein's "...long amino acid chain automatically folds into a specific stable 3D configuration...Particular protein functions depend on highly specific 3D shapes...Significant functional modification of a protein would require several simultaneous amino acid replacements of a relatively improbable nature." [19]

A protein's failure to fold correctly, risks disaster.

"The protein's most widespread role is as a catalyst in biochemical reactions, and in this role it is called an enzyme...Each reaction has its own enzyme..." which can "speed up a reaction rate by at least a million...

"An increase in rate by factors of ten billion to a hundred trillion are not uncommon...A factor of a hundred million means that what takes a thousandth of a second with the enzyme, would take about 3,000 years without it." [20]

Unlike the which-came-first-chicken-or-egg quandary, the living cell débuted on the scene, hitting the ground running with all its proteins and DNA information in place and fully functional. There's no room for a multi-million-year time gap for either one to catch up to join the other.

Enzymes--protein forms within the cytoplasm--stand guard as catalysts, expediting the life processes of the cell. Without the lightening-like speeds introduced by enzymes, biochemical reactions would take so long that they could fail to function.

The odds against the 2,000 enzymes essential to the simplest life form appearing spontaneously from inorganic matter, at one time and in one place, runs at something in the range of $10^{40,000}$ to one. [21]

Drowning in such abysmal odds, no amount of hype can transform wishful thinking into unassailable fact---even in Las Vegas.

"Amino acid, when found in nonliving material…comes in two chemically equivalent forms. Half are right-handed and half are left-handed—mirror images of each other." [22]

Amino acids exist in left-handed and right-handed formats.
Living cells build exclusively from left-handed formats.

Living cells build only from left-handed amino acids---never right-handed. The explanation escapes finite minds. The left hand/right hand amino acid mystery doomed to failure, the human attempt to artificially structure life-from-non-life.

"…Amino acids produced in Miller's apparatus were both right-and left-handed amino acids, but right-handed amino acids are poisonous to living organisms. Right-handed amino acids render proteins nonfunctional." [23]

"Amino acids in life, including plants, animals, bacteria, molds, and even viruses, are essentially all left-handed.

"No known natural process can isolate either the left-handed or the right-handed variety. The mathematical probability that chance processes could produce merely one tiny protein molecule with only left-handed amino acids is virtually zero." [24]

Just as mysteriously, amino acids revert to inorganic matter's left-handed/right-handed status at death of an organic system. Even with a laboratory-created reducing atmosphere, without a valid technique for producing life-handed amino acids exclusively, the attempt to manufacture life artificially, inevitable doom was assured.

By the year 2,000, the human genome, with its
estimated three billion+ base pairs, had been deciphered.
This complex system orchestrates 75-trillion human body cells.

Imagine every human on earth given a Rubik Cube with the challenge to simultaneously resolve the three-dimensional puzzle in sixty seconds. Then immediately be asked to correlate the colors again---also simultaneously and in sixty seconds And again…and then again, etc.

You get the picture. Impossible, of course. But that's more likely to happen than a single living cell creating itself from lifeless chemical matter.

© Chuck Nelson

Every living cell powering this elegant Canada Goose flying machine, is infinitely more complex than any mechanism ever designed by human genius.

"Recently, Eugene Koonin and others tried to calculate the bare minimum required for a living cell, and came up with a result of 256 genes. But they were doubtful whether such a hypothetical bug could survive, because such an organism could barely repair DNA damage, could no longer fine-tune the ability of its remaining genes, would lack the ability to digest complex compounds, and would need a comprehensive supply of organic nutrients in its environment." [25]

"…200 million variations, ranging from microscopic red blood cells to long, skinny nerve cells that stretch from the base of the spine to the foot…Every cell contains an estimated one billion compounds…and among these compounds are five million different kinds of proteins…

"These compounds are highly variable in shape, size, electrical charge, and configuration; many can complete a function in a millionth of a second." [26]

Tissues, organs, and systems are composed of living cells, each doing its specialized thing in synchronized concert.

"Organisms consist of a number of subsystems which are all co-adapted to react together in a coherent manner; molecules are assembled into multimolecular systems," which combine "into cells, cells into organs" resulting in a "complete organism." [27]

Heart cells code for the heart, skin cells for the skin, and brain cells for the brain. What prevents a cell from crossing over and performing the wrong service, in the wrong location, for the wrong organ?

The cell's awesome complexity in the context of the fully functional human system might have brought assumption of its accidental origin to its knees but for evolution's entrenched bias.

The dead branches of that 1837 theoretical drawing of a "tree of life" fail to explain the source of its roots and mischaracterize life's origin as an accident of nature without design, direction or purpose. The complexity of the simplest known cell only complicates attempts to build a viable theory rooted in chance.

So, where does that leave a postulate that a composite of inanimate atoms might have assembled information for living cells from nothingness, to accomplish something big beyond the competence of human thought?

The theory's genealogy lacks authenticity since all life forms are alleged to be constantly on the move, transiting to something new and different. If taken literally, not only are all species subject to extinction and replacement, life in the here and now confronts only end-of-the-line nothingness.

Seeking the source of genetic information packed into each cell's DNA would ultimately push hypothetical life by accident off the "Richter" scale of responsible scientific rhetoric.

As for genetic information, "DNA (deoxyribonucleic acid) and RNA (ribonucleic acid) molecules, which are composed of complex arrays of amino acids and are the templates for all living organisms, have yet to be artificially created." [28]

David Coppedge summarizes a wide range of biological designs mimicked by industry, recognizing that "in order to reverse engineer a system, it had to be engineered in the first place." [29]

A Madagascar spider spins silk ten times stronger than Kevlar; jellyfish are being studied in order to build a "better aquatic pump." Design of the elephant's trunk is referenced for building robotic arms. Shark skin suggests a pattern for ship hulls and swimsuits. The optical flow of honeybee eyes provides guidelines for developing navigation systems capable of complex maneuvers.

The Modern Evolutionary Synthesis has yet to be verified in laboratory tests. The chance hypothesis lacks the rationality inherent in the design inference. "The Darwinian theory of descent has not a single fact to confirm it in the realm of nature. It is not the result of scientific research, but purely the product of imagination." [30]

Awed poet, Joyce Kilmer, describing the magnificence of the humble tree, wrote insightful lines that resonate wisdom: "Poems are made by fools like me, but only God can make a tree." [31]

The poet might have added, "And only God can make a living cell."

VIII

"Nonsense of a High Order"

DNA, the *Language of Life*

"DNA contains the genetic blueprint of life...
It gives instructions to the rest of the cell to make proteins, and it
passes this same information on to the next generation ...
Without DNA, living organisms cannot survive." [1]

Carl Werner

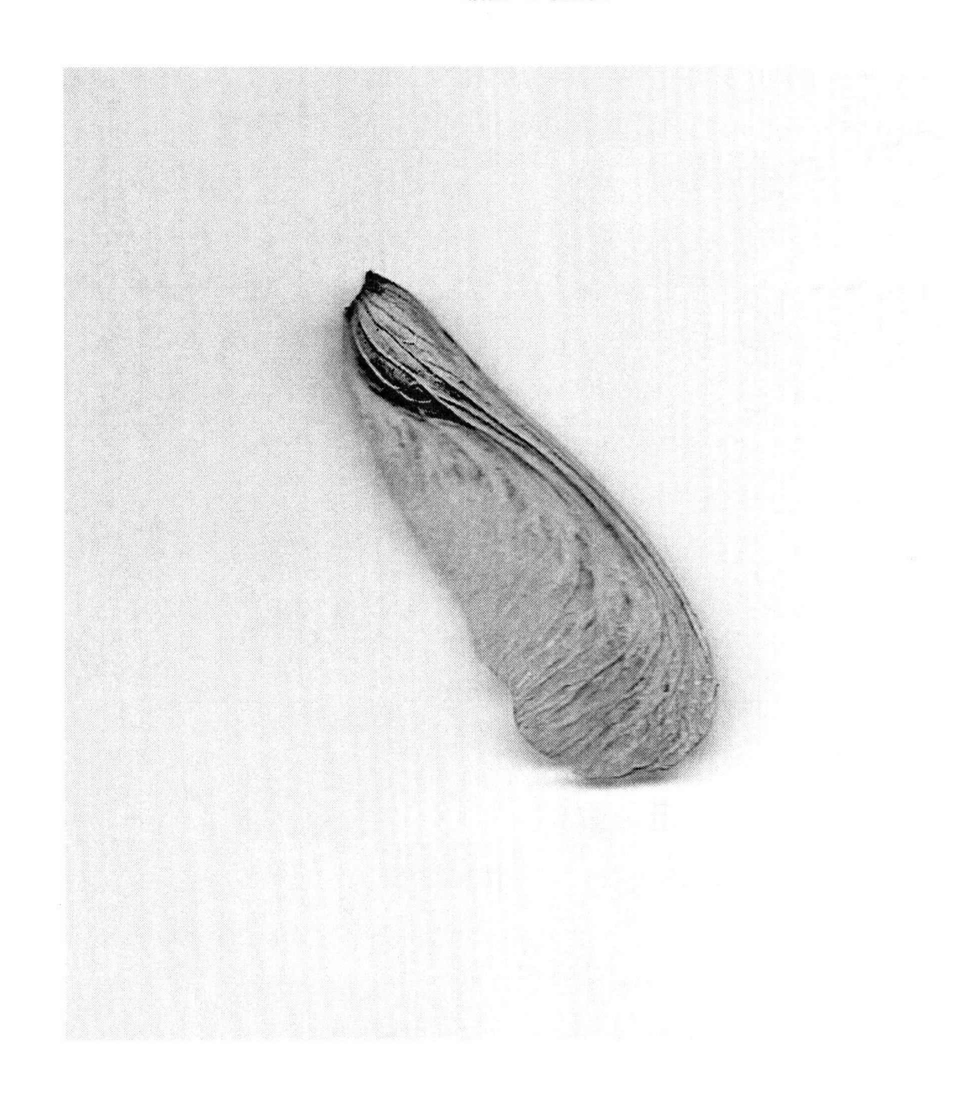

A microscopic DNA code, hidden in this "helicopter"
seed pod, dictates reproduction of a maple tree, *after its kind.*
Evolution has yet to explain the source of its genetic information.

A singe gene's DNA information can no more be created by chance than medieval alchemists could make gold by mixing recipes of other elements using abracadabra magic.

So where did a cell's information come from?

A blank computer disk offers a blank screen until some human delivers a digital code loaded with text or pictures. An electronic data bank lacks meaning or purpose without the deposit of a precisely coded message put in place by some intelligent source.

Once loaded, the coded message can be replicated *ad infinitum.*

Try playing a card game by shuffling a deck of blank cuts of white cardboard, lacking numbers or distinctive graphic prints. No winners, no losers, no game--just a meaningless shuffle of nothingness.

Or try building a bridge without the specs essential to design and transform raw steel into a graceful span over the Golden Gate. An attractive cover that binds a package of blank white paper would never make a best-seller list. It's the informative printed message that sells a book.

Matter, without information, lacks much meaning. Matter, by itself, is abstract. Unproven theory epitomizes intellectual incoherence.

First graders, up to their ears in ABC's, recognize 26 letters in the English alphabet. Kids learn to spell "cat" and "dog." By the time they reach high school, they have discovered the English language boasts a plethora of mind-bending combos of those 26 symbols---a vocabulary of significantly more than at least 200,000 English words with a capacity to communicate information on an encyclopedic scale.

Impressive capacity, but not close to matching the information combos powering the "Language of Life."

Monarchs summer in North America and then trek south annually, 2,000 miles to Mexico's warmth. Clouds of burnished gold wings, bound for their southland "resort," float at a gentle ten miles per hour, typically covering fifty miles per day. Monarchs find the identical winter home each year thanks to their own built-in GPS. The annual round-trip involves five generations---four that live only a month or so while the stalwarts that navigate the migration route to Mexico live nine months.

© Joyce Boffert

The source and complexity of the "language of life," pre-packed into a living cell, challenges imaginations.

Living ells contain definitive information blueprints, distinctive codes for every plant and animal species inhabiting earth.

Non-living chemicals exists in the form of 100+ known atomic elements. The quantity of protons contained in the nucleus of an atom gives each element its distinguishing number. Ten years after Darwin floated his ideas, Russian chemist Dmitri Mendeleev arranged the 63 then-known elements in a chart of rows and columns dubbed a *Periodic Table of the Chemical Elements.* The chart confirms chemistry's intrinsic organization.

Chemical recipes created from molecular mixes of these elements produce products underwriting industry. DNA's coded information dictates gene function with a certitude reminiscent of *The Periodic Table's* inorganic elements. The predictable results of hybridization and gene splicing contradict the feasibility of the chance hypothesis.

"What has been revealed as a result of the sequential comparisons of homologous proteins is an order as emphatic as that of the periodic table. Yet in the face of this extraordinary discovery the biological community seems content to offer explanations which are no more than apologetic tautologies." [2]

Michael Denton's, *Evolution: A Theory in Crises*, opened the door to a glimpse of molecular biology's avalanche of insight.

"Neither of the two fundamental axioms" of neo Darwinism, "...continuity of nature...linking all species together and ultimately leading back to a primeval cell" and "adaptive design...from a blind random process have been validated by one single empirical discovery or scientific advance since 1859." [3]

Even if spontaneous generation had defied impossible odds by accidentally creating a "simple," living cell, it couldn't survive, much less reproduce itself, without its own package of DNA calling the shots.

Once the assumption virus invades reasoning, superstitious nonsense follows close behind. With or without popular acclamation, assumption masquerading as science, doesn't evolve fact. Its futile to contend a cell's information code originated spontaneously from inert matter.

DNA contains the master genetic blueprint for every living organism with precision reminiscent of inorganic chemistry's Periodic Table of the Elements.

DNA exists *sui generis* (one of a kind). Multiple protein families provide an array of potential building blocks for specific living formats. Just where did that genetic information originate?

Logic slinks out science's back door when the "design inference" is ignored.

Microscopic genetic information codes every cell of every living organism with a one-of-a-kind mark.

Tucked into the cell's nucleus are the strands of deoxyribonucleic acid containing the code of life for every organism---DNA. This compendium of living data is so miniscule that it can't be seen by unassisted human sight. Organic life could not exist without an information blueprint housed in every living cell.

"Without DNA, living organisms cannot survive." [1]

Its less than rational, and certainly unscientific, to suggest DNA's microscopic strings of pre-coded information appeared accidentally after millions of years of chaotic heating, cooling, and thawing mingled with nature's rain torrents, lightening flashes, and wind gusts.

Every prototype plant and animal life on earth carries a staggering stash of unique genetic information locked in place from the get-go. DNA data banks guarantee kaleidoscopic diversity in descendant generations but never radical change to entirely new and different organisms.

The first cell's DNA base pair could not exist without imprinted language of life instructions. A single cell's DNA comes loaded with information sufficient to fill 3,000 sets of printed encyclopedias.

"...The capacity of DNA to store information vastly exceeds that of any other known system; it is so efficient that all the information needed to specify an organism as complex as man weighs less than a few thousand millionths of a gram...Each gene is a sequence of DNA about one thousand nucleotides long." [4]

These ladder-like spirals are built on side rails with alternating molecules of phosphate and deoxyribose "held together by 'rungs' called nucleotides (or bases) consisting of four specific chemical molecules: thymine (T), adenine (A), cytosine (C) and guanine (G)...Millions upon millions of nucleotides are known to exist in the nuclear DNA structure of living cells..." [5]

Living organisms use these same nucleotides but in radically different formats just as cars and skyscrapers use steel but with different designs.

Nucleotides A and T bond together to form an AT or a TA base while G and C bond as GC or CG. "A always bonds to T and the base letter G always bonds to C. Three contiguous letters on the DNA molecule (called a codon) instruct the cell to place one particular amino acid into a protein chain...

"RNA carries this template of the DNA and assembles proteins by attaching amino acids together in a chain." [6]

"Nucleotides cannot be added at will; even if they did, they could not alight themselves in a meaningful sequence....any physical change of any size, shape, or form, is strictly the result of purposeful alignment of billions of nucleotides. Nature or species do not have the capacity of rearranging them nor to add them." [7]

Sir Fred Hoyle scoffed at the concept of a genetic code, emerging from some primordial organic soup by chance, dismissing it as "nonsense of a high order." [8]

The genome's complexity and DNA source defies explanation.

Denial of DNA design, put in place at the command of an infinite intelligence, conjures up images of superstitious ancients, worshiping at the feet of inert matter, bowing in supplication before homemade idols of wood and stone.

The hexagonal honeycomb design in a beehive utilizes a minimal amount of wax to store a maximum amount of honey. A dragonfly sees through two eyes, each with 30,000 lenses. The bee gets by with 6,300.

Genes dictate the difference

© Chuck Nelson

Bird feathers reflect a rainbow of colors adding symmetric beauty to a bird's aerodynamic flying and temperature control. A microscope exposes the incredible complexity of a single human hair.

Potato bug eggs hatch in 7 days; canaries require 14; hen's eggs take 21 days; ducks and geese require 28; mallard ducks take 35; while 42 marks the timeline for the parrot and the ostrich. [10]

An elephant's four legs bend forward at the knees easing its rise to an upright position. In contrast, a horse first uses its front legs to stand while a cow starts with its hind legs.

Watermelons have an even number of stripes on the rind; oranges have an even number of segments; an ear of corn carries an even number of rows; and a stalk of wheat has an even number of grains.

And then there are banana bunches, with an even number of bananas on the lowest row, with each subsequent row decreasing by one, resulting in alternating even/odd rows for the bunch. [9]

Human genes design lachrymal glands essential to moisten eyes. Stag deer grow racks of antlers courtesy of gene code information. Thanks to distinctly unique genetic information, humans don't sprout horns on their heads and deer don't evolve tear ducts on their feet.

Just as fingerprints identify patterns unique to a person, every body cell portrays the DNA sequence of a genome, unlikely to be precisely identical with that of another human. The individual master code that dictates eye color, sex, and the shape of the face is usually one-of-a-kind.

DNA not only revolutionized investigation of life's origin but also emerged as a tool of the judicial system, shining light on crime detection and resolution. When DNA doesn't match crime scene evidence, convictions have been overturned, sometimes years after the fact.

Internet entrepreneur Bill Gates recognizes the limits of cyberspace language in contrast to the cell's ability to store and utilize living data.

"DNA is like a computer program but far more advanced than any software we've ever created." [10] It's been calculated that if "...all the copies of the DNA in all the cells of your body were straightened and laid end to end they would be about 50 billion kilometers long" [11] and would stretch from the earth to beyond the solar system.

Did the 50 billion kilometer length of microscopic strands of human DNA result from mega-millions of luck-of-the-draw mutations? Seriously?

Multiply nothingness by billions and the answer remains a vacuous zip.

"DNA and RNA molecules do not form spontaneously or abiotically in any 'primordial earth' type experiment..." [12]

Ribonucleic acid (RNA) acts as DNA's transfer agent taking the code from the nucleus to the cytoplasm. Its never been demonstrated that either DNA or RNA components of a cell can evolve spontaneously from non-living matter.

With a single rail composed of phosphate and ribose molecules and rungs with uracil (U) instead of thymine (T), messenger RNA enters the nucleus at a moment when the DNA unwinds, reads and copies the information then delivers the code to a ribosome protein-producing factory. "The order of dRNT-s of DNA determines its information content, provided that the rest of the cell's machinery is present." This refers to the "complex apparatus which duplicates DNA, transcribes it to a readable message and then reads the message and produces a functioning protein…" [13]

Not all genetic data resides in the cell's nucleus.

"…Mitochondria have their own DNA/RNA structure…a semi-independent order-giver in its own right, a computer sub-station so to speak…Human mitochondrial DNA has slightly more than 16 thousand nucleotides." [14]

If this infinitesimal information resource were placed in a teaspoon, together with DNA "necessary to specify the design" for all species of living organisms ever to have lived on the planet, "there would still be room left for all the information in every book ever written." [15]

Evolution's "RNA World Hypothesis" posits that RNA might have generated the evolution of a complete cell because it carries information and serves also as a catalyst for the manufacture of a cell's protein. This hypothesis is a dubious entry in the sweepstakes of the make-believe.

Attempting to explain the origin of first life by accident, the idea credits " 11 small carbon molecules…that could have played a role in other chemical reactions that led to the development of such biomolecules as amino acids, lipids, sugars, and eventually some kind of genetic molecule such as RNA." [16]

Little did those infinitesimal "molecules" comprehend the magnitude of the creative authority theoretically vested in their atomic composition.

The language of life can't exist without proteins, and proteins can't exist without the molecule that provides and stores instructional information. RNA cannot survive on its own without being an essential part of the complete living organism---nor can it account for the source of its original genetic information. DNA, RNA, proteins, amino acids, and cell membranes exist as a composite, mutually interdependent whole.

It's an all or nothing, a package deal.

Its axiomatic that a functional cell requires a precise dose of genetic directions to tell its amino acids and proteins just how to function as a one-of-a-kind, reproducing organism.

The genomes of different species show vast differences that defy bridging by simply reshuffling the card deck and re-dealing the same genetic information. Any change is limited to the potential combinations of the original 52 cards in the deck unless different cards are added to the mix with entirely new and different information.

The genome of a chimpanzee doesn't match the genome of a human. Nor is there a whit of evidence that an elephant, a giraffe, a butterfly, or a giant Sequoia could have descended from a common, single-cell ancestor.

Amino acids don't organize themselves without a genetic code orchestrating the process. Dean Overman discounts "biochemical predestination" in molecular genetics.

"The enormous information content of even the simplest living systems...cannot in our view be generated by what are often called 'natural' processes...

"For life to have originated on the Earth it would be necessary that quite explicit instruction should have been provided for its assembly...There is no way in which we can expect to avoid the need for information, no way in which we can simply get by with a bigger and better organic soup..." [17]

It's intellectually incoherent to assert life and its built-in code of information appeared spontaneously, by chance, independent of intelligent cause. Words and layouts in a news feature don't compose and release themselves without design or designer.

Like the machine with the sole function of turning itself off, human minds have proposed the inherently impossible--the creation of intelligent information from a non-intelligent source.

"...Since most proteins are, on average, 300 amino acids long, the DNA needed to make just one protein would have to be approximately 900 letters long." And since the first living organism would have required 20 or so basic proteins to function, each with 900 letters of DNA, "...18,000 letters of DNA would be needed to theoretically form the first single-cell organism." [18]

Now factor in the length mandate.

"Scientists have observed DNA strands forming naturally in the laboratory, but only in strands of up to 20 letters in length...

"After 20 letters of DNA come together, the DNA begins to break apart. Simply put, long strands of DNA (hundreds of tens of thousands of letters long) do not form naturally because chemical properties of DNA prevent this." [19]

Then there's the required sequence pattern.

"The 18,000 DNA letters have to be lined up in a particular order to call for particular amino acids in a particular set of proteins for life to begin...The odds of winning the national Powerball Lottery every day for 365 days are 1/4,244 followed by 2,881 zeros...

"The chances of DNA forming spontaneously with the proper letter sequence" requires 10,837 zeros.[20] Calculating the odds requires recognition of the number of genes to be reckoned with.

Mycoplasma genitalium, with its 482 genes and 580,000 bases, may be the "simplest known self-reproducing life form." [21] The genome of a mammal contains a string of "from two to four million" symbols "...that would fill two thousand volumes---enough to take up a library shelf the length of a football field. All this is in the tiny chromosomes of each cell." [22]

The mathematical odds against a single cell appearing by accident push computations over the edge. The already impossible eventually confronts the ultimate challenge of merging millions of cells into a cohesive living unit composed of a functioning conglomerate of tissues, organs, and systems.

Based on mathematical probability factors alone, "...any viable DNA strand having over 84 nucleotides cannot be the result of haphazard mutations. At that stage, the probabilities are 1 in 4.80 x 10^{50}...Mathematicians agree that any requisite number beyond 10^{50} has, statistically, a zero probability of occurrence...

"Any species known to us, including the smallest single-cell bacteria, have enormously larger numbers of nucleotides than 100 or 1000...This means, that there is no mathematical probability whatever for any known species to have been the product of a random occurrence---random mutations..."[23]

However contrived the calculation, *impossible* spells impossible.

"...The DNA molecule may be the one and only perfect solution to the twin problems of information storage and duplication for self-replicating automata...

"It is astonishing...that this remarkable piece of machinery which possesses the ultimate capacity to construct every living thing that ever existed on Earth, from a giant redwood to the human brain, can construct all its own components in a matter of minutes and weigh less than 10^{-16} grams. This "...is of the order of several thousand million-million times smaller than the smallest piece of functional machinery ever constructed by man." [24]

"The difference in DNA between species resides almost strictly in the sequential positioning of the nucleotide...No two individual plants or animals have DNA spirals that are identical..." [25]

"Bacterial genomes are very different from eukaryotic genomes [Bacteria are Prokaryotes, single-cell life forms with an open internal cell structure while Eukaryotes, also single celled, occupy multiple compartments] in that they usually do not possess 'exons' and 'introns' and do not have extensive scaffolding. Bacteria even lack membrane-bound nuclei, so that their genes can be expressed much more quickly than those of the eukaryotes..." [26]

"...Pure unguided random events cannot achieve any sort of interesting, or complex end...The fact remains that nature has not been reduced to the continuum that the Darwinian model demands, nor has the credibility of chance as the creative agent of life been secured." [27]

To qualify as anything more than a "fairy tale for grown-ups," evolution theory needs confirmation at the molecular level.

A congenitally flawed idea crafted from ambiguous abstraction lacks substance. Molecular biology does nothing to demonstrate the genetic continuity critical to the corroboration of evolution theory---but it does confirm discontinuity.

"At a molecular level…there is no trace of the evolutionary transition from fish to amphibian to reptile to mammal. So amphibia, always traditionally considered intermediate between fish and the other terrestrial vertebrates, are in molecular terms as far from fish as any group of reptiles or mammals." [28]

Primitive minds, tainted by superstition, studied what they could see—embryos and bones. Without access to the all-seeing eye of the electron microscope, similar appearance could be misinterpreted as relatedness. The world might have been spared a library of specious speculations had molecular biology emerged as a nineteenth-century science.

It's universally recognized "…physical growth is the result of a very specific sequence of hundreds of thousands of nucleotides in its own DNA" which dictates the order giving sequence that "seeps down to the growth mechanisms. It never acts in the reverse direction…" [29]

Lacking evidence of a convincing pattern of transitional life forms after millions of fossil discoveries, evolution theory lives or dies in DNA's molecular world. But the microscope doesn't oblige.

So, back to the drawing board; just where did that DNA information stashed in DNA originate? Certainly DNA didn't creep into the world playing roulette with nature's chemistry set.

An intelligent source is the rational option.

Michael Denton confirms the inability of chance to produce change necessary to close the massive gaps existing at the molecular level.

"No evolutionary biologist has ever produced any quantitative proof that the designs of nature are in fact within the reach of chance…It is surely a little premature to claim that random processes could have assembled mosquitoes and elephants when we still have to determine the actual probability of the discovery by chance of one single functional protein molecule." [30]

Evolution theory hangs by a tattered thread, trapped by the litany of unproven assumptions its sponsors spin to the public. With a giant zip on the science scoreboard how does the idea survive?

What keeps a bankrupt franchise in business?

Evolution's devalued life concept exists exclusively on paper. Little else remains for the fragments of Darwin's dream but an overdue eulogy. The core question deserves repetition: *just where did the information, pre-loaded into the cell's DNA, come from?*

Objectivity directs the intellect to a *design inference* answer!

In a futile quest for missing transitional forms, evolutionary theory has lost its academic battle in fossil cemetery fields and molecular biology laboratories. Instead of offering salvation for a decrepit idea, molecular discontinuity is writing the theory's overdue obituary.

Early in the twentieth century, the ball was passed along to mutations in a last, desperate effort to salvage flawed fiction.

Antony Flew, at one time one of the world's foremost atheistic philosophers. observes: "The only satisfactory explanation for the origin of such 'end-directed, self-replicating' life as we see on earth is an infinitely intelligent Mind" [31]

IX

"Refuse Material of Nature's Workshop"

Mutations

"No matter how numerous they may be, mutations do not produce any kind of evolution…There is no law against day dreaming, but science must not indulge in it." [1]

Pierre-Paul Grassé

© Ng Wei Keon

Does a peacock owe its array of feather designs to a series of lucky mutations?

Even if mutations could deliver new information to the genome, this doesn't explain the source of a cell's original genetic information.

Genetic mutations typically corrupt DNA codes, degrading the genome, and contributing nothing to the "evolution" of different life kinds.

When Darwin first floated his chance hypothesis, he was as ignorant of DNA as he was of a flash drive storing computer information. But once having shared his conjecture with the public, he may have felt compelled to offer an explanation as just how incremental transitions occurred.

Already out on a limb with his *tree of life*, he latched on to a dubious "solution:" why couldn't physical change, activated by the use or disuse of a body part, be passed along to a descendant?

Differentiating between *somatic* and *germ* cells, he trotted out gemmules as the mechanism that would preserve and pass along acquired physical traits.

The nonexistent gemmule is defined as "a hypothetical particle of heredity postulated to be the mediating factor in the production of new cells in the theory of pangenesis." [2]

Bit-by-bit, gradual changes supposedly accumulated en route to an entirely new and different organism. The innovative naturalist seized upon the towering neck of the giraffe, as prime evidence of the process.

Fairy tale logic manufactured the fiction that when persistent droughts dried the fields of grass grazed by giraffe ancestors, survival hinged on stretching their necks toward the sky to nibble tree leaves.

Theoretically, the animal's bone and muscle structure preserved those neck stretches and passed each miniscule change to the next generation, courtesy of gemmules. Over time, these modest adaptations supposedly evolved the long-necked giraffe that populates African plains.

Gemmules existed only in Darwin's mind. Never observed in nature, the far-fetched myth made no sense. It would be early into the twentieth century before reality intruded and the world would learn that gemmules were not the magic elixir providing natural selection something to select.

As to the neck of an 18-foot-tall giraffe with a heart 2 ½ feet long, mutations contribute nothing more to explain its origin than do make-believe gemmules.

The heart strong enough to pump blood up a giraffe's neck to the brain is also powerful enough "to burst the blood vessels of its brain" when it reaches down for a drink of water. But when the giraffe bends down, "a protective mechanism" kicks in causing "valves in the arteries in its neck" to begin to close. [3]

Did Darwin wonder why other animals, such as zebras, wouldn't also evolve longer necks? Or why a human muscle builder couldn't pass along bulging, exercise-built muscles to his children? Either he hadn't heard or had ignored the insightful news from Mendel's garden, released in 1865,

The eloquence filtering through the stentorian tones of Darwin's persuasive English accent proved susceptible to superstition, tradition and personal bias prior to access to modern technology tools.

Theories of origin came cluttered with the same medieval hot air that floated spontaneous generation fallacy. Oblivious to genes and DNA, evolutionists went looking in the wrong direction for theory confirmation, relying on spin the bottle luck and mega years of evolutionary gestation. Evidence confirming a continuous chain, linking simple-to-complex life forms in fossil fields, lay in discontinuous disarray.

Not until the advent of molecular biology did discovery doors swing open, inviting comprehension. The world might have been spared specious speculations had molecular biology been a nineteenth-century science.

Gregor Mendel (1822-1884), an obscure Czech monk and younger Darwin contemporary, discovered fundamental genetic principles while he was working in the quietude of a monastery garden shortly after the English naturalist went public with his postulated theory.

When the twentieth century dawned, Gregor Mendel's garden set the stage for a direct assault on the Darwinian notion that enamored nineteenth century academics.

Mendel bred several generations of garden peas, focusing on the plant's seven basic characteristics. What he discovered about hybrids clashed head-on with Darwinian conjecture. Mendel's experiment revolutionized knowledge of the mechanism of inheritance.

Wikipedia

Gregor Johann Mendel's (1822-1884) discovery of the Law of Genetics threw a figurative monkey wrench into evolution's iconic mechanism.

Mendel read his findings before the *Brünn Society for the Study of Natural Science* in 1865. His eyebrow raising report appeared in the Society's *Journal* in 1866, warranting distribution to 120 libraries, including some in England

and eleven in the United States. His independent scholarship, otherwise ignored or overlooked at the time, earned reference in the *Encyclopedia Britannica's* 1892 edition.[4]

The landmark findings, once obscure and virtually unnoticed, were translated eventually and published in English in 1900. Pushing public release was 39-year-old British biologist, William Bateson (1861-1926), an admirer of Mendel's scholarship, founder of the science of genetics and later president of the British Association for the Advancement of Science.

Publication rocked academic circles.

There is speculation as to whether Darwin's theory of evolution would have "evolved" and seen the light of day had he studied Mendel's findings. Bateson, a British biologist, expressed doubts. "Darwin would never have written the *Origin of Species* if he had known Mendel's work." [5]

Maybe, but we'll never know.

What we do know is that today's cell knowledge, with discovery of its DNA strands, correlates with Mendel's law of heredity and does nothing to corroborate Darwin's theory.

Darwin's co-evolutionist, Alfred Russell Wallace (1823-1913), viewed Mendel's findings with alarm, posing a serious threat to evolution. He dismissed Mendel's discovery and discounted *mutations* as minimally significant. "On the general relation of Mendelism to evolution, I have come to a very definite conclusion. That is, that it is really antagonistic to evolution." [6]

He had reason for concern!

"As playing any essential part in the scheme of organic development, the phenomena seem to me to be of the very slightest importance. They arise out of what are essentially abnormalities, whether called varieties, 'mutations,' or sports." [7]

He saw "...their extinction under natural conditions more certain and more rapid, thus preventing the injurious effects that might result from their competing with the normal form while undergoing slow adaptive modification...

"Any species which gave birth to a large number of such abnormal and unchangeable individuals would be so hampered by them whenever adaptive modification became necessary that the whole species might be in danger of extinction."

Wallace scorned these abnormalities as "refuse material of nature's workshop, as proved by the fact that none of them ever maintain themselves in a state of nature." [7]

Presumably unforeseen by Wallace, his dismissal of mutations as mere "refuse material," ran diametrically counter to the subsequently postulated *Modern Synthetic Theory of Evolution*! He didn't live long enough to learn that what he described correctly as "abnormalities" would be seized upon eventually as the Holy Grail in an attempt to salvage a crumbling tradition.

Mendel's pioneer research launched the science of genetics, confirming life forms function through precisely expressed information codes with predictable results. His law of genetic inheritance blew the cover off empty rhetoric embellishing make-believe. The discovery opened doors to genes, chromosomes, and DNA.

"Cell biologists identified chromosomes as the carriers of Mendel's heredity factors, and in 1909 Wilhelm Johanssen named them 'genes.'" [8]

What Mendel did with garden peas, Hugo deVries confirmed with the primrose. He reported flowers rising "...suddenly, spontaneously, by steps, by jumps. They jumped out among the offspring." DeVries described these hybrid variables as "new species" which he labeled "mutations." [9]

The word was out: Given time, "mutations" were incorporated as key components of evolution's iconic lexicon.

Wallace was not the only influential evolutionist reluctant to jump for joy when confronted with Mendel's law of genetic inheritance. The news shook the faith of others with seismic impact.

Princeton's Prof. Scott complained that the findings "...rendered but little assistance in making the evolution process more intelligent, but instead of removing difficulties have rather increased them." [10]

A shaken University of Paris Chair of Evolution professor, voiced alarm to a 1916 Harvard audience: "It comes to pass that some biologists of the greatest authority in the study of Mendelian principles of heredity are led to the expression of ideas which would almost take us back to creationism…The data of Mendelism embarrasses us quite considerably." [11]

Zoology Professor, E.W. McBride reminisced for *Science Progress* the year of the 1925 Scopes Trial. He recognized Mendel's Law as a potential wet blanket cast over evolutionary theory after first being greeted with "enthusiasm."

"We thought at last the key to evolution had been discovered.

"But as our knowledge of the facts grew, the difficulty of using Mendelian phenomena to explain evolution became apparent, and this early hope sickened and died. The way that Mendel pointed seemed to lead into a cul-de-sac." [12]

His "cul-de-sac" analysis summarized the dilemma.

Its more than a quantum leap to suggest a string of DNA from human cells stretching 50 billion kilometers into space could be built by multi-millions of random chance quirks of nature flowing seamlessly from earth's first ever life form. The Mendel effect reverberated far beyond garden peas. Evolutionists recognized the sheer absurdity of gemmule poppycock.

The daunting revisionism task was handed to Wallace's "refuse material."

Mendel's genetics law eventually caught the attention of scholars and crept into science vocabulary early in the 20th century.

The stage beckoned for a chance hypothesis upgrade that would rally the faithful. Devout evolutionists scrambled for alternatives for a discredited theory facing collapse on academic ropes.

And not a moment too soon! Crescendos of disgruntled voices could be heard reacting negatively to serious threats to an unproven tradition.

"[T]he theory suffers from grave defects, which are becoming more and more apparent as time advances. It can no longer square with practical scientific knowledge, nor does it suffice for our theoretical grasp of the facts...

"No one can demonstrate that the limits of a species have ever been passed. These are the Rubicons, which evolutionists cannot cross. Darwin ransacked other spheres of practical research work for ideas...but his whole resulting scheme remains, to this day, foreign to scientifically established zoology...actual changes of species by such means are still unknown." [13]

Evolution has shown innovative flexibility when its survival is at risk. During the 1859 to 1872 interval, Darwin repeatedly modified his *Origin of Species* manuscript, ultimately increasing the original 150,000-word text to 190,000 words.[14]

Responding to Mendel's law, evolutionists wracked their collective brains for some fashionable alternative to the discredited idea that acquired physical characteristics could be inherited. Rather than a once-perceived nemesis to a flawed idea, mutations emerged as the cure-all designed to redeem an endangered idea from the slagheap of faux science.

Theodosius Dobzhansky floated the mutation life preserver in 1937, when he celebrated the transformation of Wallace's "refuse material" into the exalted role of evolution's "raw material." He postulated that "mutations and chromosomal changes...constantly and unremittingly supply the raw materials for evolution." [15]

Not only do mutations corrupt genetic codes, they can kill.

In 1941, evolutionist movers and shakers put their heads together in an effort to reinvent Darwinian thought by blazing a trail through the quick sands of genetic mistakes.

Fleeing the gemmule pathway, loyal revisionists seized Dobzhansky's straw in the wind, and used academic sleight-of-hand to redefine "refuse material" to read the "raw material." available to partner with *natural selection*, thereby rewriting the mechanism powering Darwin's dream. This legerdemain turned the looming threat of the mutation negative into a newly minted positive.

Designating mutations as the raw materials for evolution seems as fanciful as envisioning the equivalent of a thundering Niagara Falls spraying its mists in a Mojave Desert mirage. Duly revived, the now updated dogma was awarded the lofty title, *Modern Synthetic Theory of Evolution*, and reintroduced to the world with fanfare's bells and whistles.

© Oliver Hoffmann

Hybrid tinkering, not mutations, generates dazzling variety, but the end product remains poinsettias.

Faithful Darwinists, symbolically walking in "tall cotton," met in Chicago in 1959 to celebrate the centennial anniversary of the publication of *The Origin of Species*.

Sir Julian Huxley, grandson of the 19th century Huxley who ran public interference for Darwin's original thought, made no apologies for the patchwork massaging of Darwin's grossly incorrect handiwork.

Waxing eloquent, Sir Julian testified to evolution's authenticity.

He assured celebrants that "Darwin's theory...is no longer a theory but a fact...We are no longer having to bother about establishing the fact of evolution..." [16]

In a spate of bravado, Huxley ruled out "either need or room for the supernatural..." [16]

Despite the exuberant chest thumping, the revised synthetic theory drew cheers mixed with some jeers. Early in this century more than 500 scientists signed-on to a declaration expressing reservations.

"We are skeptical of claims for the ability of random mutation and natural selection to account for the complexity of life. Careful examination of the evidence for Darwinian theory should be encouraged." [17]

Not only are mutations typically synonymous with genetic degradation, no valid evidence demonstrates that mutations have ever successfully provided the raw information essential to transit one living organism to some new and different life kind.

Redundant clichés citing finch beaks, fruit flies, and bacteria modifications as evidence of evolution confirm only the versatility potential built into a preexisting genetic code. Genetic chasms, separating distinctly different kinds of organisms, have yet to be bridged by the conjectured mutation/natural selection combo.

No single mutation, or series of mutations, collaborating with natural selection, can activate random chance transformation to a new life kind without the addition of new genetic information.

Mutations paired with natural selection may shift a descendant organism laterally or down but never vertically up the genetic staircase to an entirely different kind of *family, order, class* or *phyla.*

"Mutations take place, but they are either reversible, deteriorative, or neutral...If one must depend on mutation and natural selection to produce new species---let alone, new families, orders and phyla as evolutionists assume, then not even billions of years would suffice." [18]

"Whoever thinks macroevolution can be made by mutations that lose information is like the merchant who lost a little money on every sale but thought he could make it up on volume." [19]

Incapable of introducing new information, mutations degrade overall fitness, typically impairing what's already there. Rather than a silver bullet to salvage a bankrupt theory, mutations have yet to find a niche in the pantheon of respectable science. Mutations typically degrade fitness.

Natural selection is genuine, but don't hold your breath expecting evolutionary change if natural selection relies on mutations. Assuming mutations provide the "raw material" for natural selection to create new and different living organisms makes no more sense than believing an improved computer evolves from a virus-infected hard drive.

© Mayskyphoto

Can a series of chance mutations account for differences between colorful macaws and regal peacocks?

Natural selection works, counter to the other half of evolution's equation, by screening out harmful mutations. Thanks to repair enzymes, mutations are subject to an organism's self-correcting mechanism. If the repair system itself mutates, the organism's survival could be jeopardized. Mutations are much too slow to accomplish evolution's conjectured mission.

"Natural selection can serve only to 'weed out' those mutations that are harmful, at best preserving the 'status quo.' " [20]

"Natural selection can act only on those biologic properties that already exist; it cannot create properties in order to meet adaptational needs." [21]

"No one has ever produced a species by mechanisms of natural selection." [22]

Natural selection alone, offers nothing more than tautology's circular reasoning.

"...The Darwinian theory of natural selection, whether or not coupled with Mendelism, is false." [23]

X

Gregor Mendel's Monkey Wrench

Sand in the Gears

"Major functional disorders in humans, animals and plants are caused by the loss or displacement of a single DNA molecule, or even a single nucleotide within that molecule...There is no evidence for beneficial...genetic mutations." [1]

Richard Milton

© Ilia Shalamaev

Can thousands of random mutations account for a kingfisher's superb fishing skill?

"...Mutations are rare events. Any particular new DNA mutation will occur only once in about 100 million gametes." [2]

Mendel's Law of Genetics demonstrates precision in the world of organisms comparable to the Table of the Elements in the inorganic.

"When a single mutation occurs in a single newborn, even if it is a favorable mutation, there is a fair probability that it will not be presented in the next generation because its single carrier may not, by chance, pass it on to its few offspring." [2]

Change in a single nucleotide would be the smallest possible modification of a genome. One evolutionist guesstimated that as few as 500 mutations could evolve a new species. No matter how many billions of chances and multi-millions of years allocated, the chance of those 500 allegedly beneficial mutation steps succeeding continues impossible ad infinitum. Mathematic odds challenge that possibility.

"It's a matter of chance that a mutant survives. It might spread through the population and take it over, but more likely it will just vanish...even good mutations are likely to disappear from the population.

"The chance of 500 of these steps succeeding is 1/300,000 multiplied by itself 500 times. The odds against that happening are about $3.6 \times 10^{2,738}$ to one, or the chance of it happening is about $2.7 \times 10^{-2,739}$...It's more than 2,000 orders of magnitude smaller than...impossible." [3]

Computations of the likelihood for any single event beyond 10^{50} chances qualifies as impossible, unless absolutism is willing to concede the intrusion of a miracle.

The impossible trumps the improbable.

What are the odds of six billion human dice throwers coming up with identical number sets at the precise nanosecond of time in 4.6 billion years, assuming each player could live so long?

The odds stretch plausibility. Reinventing evolution by reliance on mutations as the theory's "raw materials" is fanciful. Genetic code corruption by mutation doesn't provide evolution's "raw material" but does contribute to deterioration of the genome."The genome is fragile, subject to environmental insults (radiation, oxidation etc.) Extensive repair systems are guarding the integrity of the genome second-by-second. Without their work, life would become extinct." [4]

Desert mirages don't deliver sparkling water. A single mathematical miscalculation can disrupt a rocket's moon shot. Mistakes in pharmaceutical products can kill.

Suggesting mutations evolve new life kinds makes no more sense than claiming disease promotes good health.

Beautiful, vivacious, four-year-old Kylie McPeak appeared to be every parent's dream of an ideal daughter.

But then, her mom sensed a "feeling" that something was wrong.

The vigilant mom had reasons to worry. By the time Kylie was six, the twitching became an uncontrollable writhing and her sweet child's voice began to quaver. Stumped local doctors encouraged the parents to present Kylie's case to the National Institute of Health's Undiagnosed Diseases Program. [5] Baffled for a time, UDH eventually spotted the culprit--a mutated gene present on human chromosome 6 that codes for the protein laforin, usually a heart-wrenching death sentence for the child victim. The genetic mutation that triggers Lafora disease in kids is a poor candidate as evidence that mutations provide evolution's "raw material,"

A growing roster of serious mutation-caused genetic disorders, including the dreaded HIV curse, had been identified by the close of the twentieth century.

So what's going on? Where is the new genetic information essential for natural selection to select and to do its thing?

Ultraviolet light, nuclear radiation, and those pesky virus parasites rank as prime candidates for mutating genetic codes. The virus culprit lurks as a prime villain. Just as electronic viruses wreak computer havoc, organic viruses lead a parade of enemies inflicting devastating damage to a genome. Evidence that virus parasites appear to be anything but life-friendly continues to pile up.

In recent years, millions of honeybees in the United States have died, mystifying science until 2010. It is now believed the twin culprits may be the insect iridescent virus attacking bees inflicted with the *Nosema ceranae* fungus [6] The havoc wreaking CCD (Colony Collapse Disorder) contributes nothing positive in support of evolutionary change.

HIV, the human immunodeficiency virus that causes AIDS, is a retrovirus. Retroviruses are viruses composed of RNA. They contain an enzyme that gives them the unique property of transcribing RNA (their RNA) into DNA. The retroviral DNA can then integrate into the chromosomal DNA of the host cell to be expressed there.

Its axiomatic: lacking ability to reproduce itself without access to a living, host cell, a virus doesn't qualify as a candidate for first life. Since the parasite cannot reproduce without latching onto a pre-existing "host cell," the how, when, and where the insidious "mini-monster" first intruded on already-existing life confounds.

The origin of a virus mystifies. The havoc it inflicts is obvious. The pesky parasite can mutate both DNA and RNA with dire consequences.

"The viral DNA or RNA then takes control of the cell's functions, including reproduction. The host cell has no choice but to make copies of the virus until the cell explodes, sending hundreds of viruses out to other healthy cells." [7]

Non-mutational changes appearance (non-genetic factors that don't modify underlying DNA sequence) may change gene expression that can alter appearance.

"The modification of epigenetic features associated with a region of DNA allows organisms, on a multigenerational time scale, to switch between phenotypes that express and repress that particular gene...

"...Most of these multigenerational epigenetic traits are gradually lost over several generations...When the DNA sequence of the region is not mutated, this change is reversible." [8]

Reversible modifications of gene expression do not change the underlying DNA sequence of an organism.

As to the tired cliché that mutations benefit bacteria by building immunity to antibiotics such as streptomycin, reality indicates "the mutation reduces the specificity of the ribosome protein, and that means losing genetic information" with "a loss of sensitivity." Despite any "selective value," this mutation "decreases rather than increases genetic information." [9]

"...Antibiotic resistance is the result of loss of a protein, loss of the binding capacity of a protein, or the loss of a transporting system...It's a loss of something...If you're removing a transport protein to eliminate the bacteria's sensitivity to antibiotics, then how is that explaining common descent by modification...?" [10]

"There aren't any known, clear, examples of a mutation that has added information." Rather, mutations lead "to a loss of sensitivity to the drug...the effect is heritable, and a whole strain of resistant bacteria can arise from the mutation...A change in one of its proteins is then likely to degrade the organism...

"Information cannot be built up by mutations that lose it." [11]

"For information to build up in living organisms, it must be created somewhere... Although in some special cases a loss of information can lead to an advantage for the organism, the large-scale evolution for which the

NDT [neo-Darwinian Theory] is supposed to account cannot be based on such mutations." [12]

"Most mutations which cause changes in the amino acid sequence of proteins tend to damage function to a greater or lesser degree...most of the amino acids in the centre of the protein cannot be changed without having drastic deleterious effects on the stability and function of the molecule." [13]

"There are all kinds of mutations that eliminate proteins. They may eliminate transport protein, an enzyme, the action of an enzyme, or regulatory systems." [14]

Mutations are "not making new transport proteins. They're not making new regulatory systems! Antibiotic resistance is an example .

"Every time you read about antibiotic resistance...they're going to talk about this as...an absolute example of evolution. There's no mutation that gives them [evolutionists] what is necessary for common descent with modifications." [14]

One of Darwin's false assumptions "was that natural selection had a building or creating capacity, and it doesn't." It removes information.

"Every molecular example of a mutation we currently have fails to provide a mechanism than can account for the origin of any genetic activity or function." [14]

Crippling mutations respect no human. Mutations unload a bleak litany of physical flaws and debilitating diseases on the genome. Harmful gene mutations have plagued humans with a list of 4,500 already identified bad results—and still counting.

Since 1990, "discoveries of heart-handicapping mutations have been pouring out of numerous labs at an ever-increasing rate, yielding more than 100 mutations in more than a dozen genes." [15]

"A genetic polymorphism called 11307K in either of...two APC genes doubles the risk of colon cancer." [16]

A monster of a man, twisted and bent almost in half before his premature death, tottered feebly, aided by a walker. Parkinson's disease, an insidious scourge, laid low the robust physique of the former football star. A mutated gene stands accused as the culprit. [17]

Progressive myoclonus epilepsy is caused by a gene mutation on chromosome 21. Treacher Callins Syndrome, hemochromatosis, is linked to a defective gene on chromosome 6.

A caring, elementary school teacher, was forced to cope with Acromegaly's brutal disfigurement and the bullying taunts of the mean spirited; the disease came from an acquired, genetic mutation.

A gene mutation on chromosome 5 generates deformities of the face, ears, down-slanting eyes, and deafness.

A chromosome 9-gene mutation causes skin cancer.

A mutated gene on chromosome 16 is tied to fanconi anemia, affecting children who rarely live past their sixteenth birthday.

A gene missing from chromosome 7 causes Williams Syndrome.

Anhidrotic ectodermal dysplasia, caused by a mutation on the X chromosome, can afflict victims with baldness, loss of teeth, or deprive them of the ability to sweat. [18]

"...Not one mutation that increased the efficiency of a genetically coded human protein has been found.

"Instead of a 'blind watchmaker,' the mutations behave like a 'blind gunman,' a destroyer who shoots his deadly 'bullets' randomly into beautifully designed models of living molecular machinery. Sometimes the 'bullets' only cause minor damage; sometimes they maim and cripple; sometimes they kill." [719]

There are "genetic flaws that make people fat..." [20] Werner's syndrome results from a mutated site on human chromosome 8 causing victims to "age prematurely fast and usually die before they reach 50." [21]

Yet another genetic mutation "...causes children to die of old age...Children with Hutchinson-Gilford progeria syndrome age at a rate five to 10 times faster than normal. They lose their hair, their skin wrinkles, and they die of arteriosclerosis, or hardening of the arteries, by their early teens." The defect is in "...a gene that controls the structure of the nucleus..." [22]

Progressive blindness, described as retinitis pigmentosa, is a disorder linked to a gene from the X chromosome. "Best's macular dystrophy...destroys the part of the retina responsible for the sharpest vision 'has been linked to mutations' in the gene now called bestrophin." [23]

Spontaneous blood clots can form with the power to cause sudden death where a "patient with the disorder has inherited at least one defective gene encoding protein C." [24]

A mutated gene on chromosome 11 contributes to inherited hearing impairment. [25]

Then there's the smorgasbord of birth defects caused by mutated genes: muscular dystrophy, spinabifida, cystic fibrosis, Huntington's Disease, hermachromatosis, and Down's Syndrome.

Leukemia may result when a piece of chromosome moves to another chromosome in the midst of cell division.

Regarding humans, researchers "calculated an unusually high rate of 4.2 mutations per generation, of which 1.6 diminish the fitness of the species...the species must survive in part because people who have accumulated dangerous mutations are least likely to successfully have children...

"The human reproductive strategy helps purge harmful mutations in batches...they mix their genes with another's, and presumably some of the worst defects aren't passed along. That wouldn't happen if humans reproduced asexually." [26]

Mutated human genes labeled APOE, CLU, and PICALM have been targeted as likely culprits inflicting Alzheimer's disease on the lives of five million Americans. [27] "Meat contaminated by a virus that can be lethal is being sold to consumers -- creating a public health threat that has largely flown under the the radar due to powerful industry interests and lax accountability at the federal agency in charge of ensuring food safety, according to recent studies and a prominent investigative journalist.

"'It makes salmonella look like a picnic,' is how David Kirby, an investigative journalist who has written about MRSA, a life-threatening pathogen, described it in an interview with Consumer Ally.

"MRSA (Methicillin-resistant Staphylococcus aureus) is an antibiotic-resistant staph infection that kills about 20,000 Americans -- more than the number of people who die from AIDS -- each year." [28]

DNA is so delicate that one report suggests: "Every exposure to tobacco, from occasional smoking or secondhand smoke, can damage DNA in ways that lead to cancer." [29]

Inevitably, evolution's mutation "cure-all" itself will suffer further from still-to-come discoveries, certain to expose virus-caused-mutations for the culprits they are--debilitating genetic mistakes.

The glaring exception to the universal commonality of the science of order and the mathematically real is evolution's mutant "science" where the logic of the measurable equation is discarded in favor of mutation's assumptive myth.

Evolution ignores the real and postulates the imaginary. Empty rhetoric lacks the mantle of substance once all cosmetic hype, bells, and whistles are taken off the table. Slick diagrams, catchy slogans and colorful imagination can't guarantee scientific respectability. Against impossible odds, evolution attracts minds to the obscenity that an ancestral, grand pappy fish spawned descendant humanity using mutations harnessed by the chance hypothesis.

Truth stands tall in three dimensions; it needs no defense. Falsehood collapses on itself, melting to nothingness when exposed to the scrutiny of hard evidence under the penetrating rays of discovery's noonday sun.

Physician Sean Pitman articulates the reality that the fossil record doesn't rank with Mendel's Law of Genetics in evaluating evolution.

"...Interpretations of fossils and the geologic column is not as definitively precise and conclusive as dealing with genetics and the Darwinian mechanism..." [30]

Pitman charges that "if Darwinian-style evolution happens or doesn't happen, it happens or doesn't happen genetically...The Darwinian mechanism of random mutations combined with natural selection was statistically untenable – dramatically so. Given billions or even trillions of years of time, it was hopelessly inadequate to explain the origin of novel functional biological information beyond very very low levels of functional complexity… I was especially shocked at the use, by modern scientists, of Mendelian variation as a basis for Darwinian-style evolution over time.

"Mendelian variation isn't evolution at all. It is simply a difference in expression of the same underlying gene pool of options where the gene pool itself doesn't change." [30]

Louis Bounoure's blunt analysis concludes that evolution "is a fairy tale for grown-ups. This theory has helped nothing in the progress of science. It is useless." [31]

Ongoing discoveries in the molecular world suggest mutations may prove to be Mendel's monkey wrench, undermining evolution theory by spilling sand into genetic gears.

The late Stephen J. Gould, articulate Harvard biologist, undercut the core essence of evolution's *synthetic theory* with cold-eyed logic: "You don't," he said, "make new species by mutating the species…

"A mutation is not the cause of evolutionary change." [32]

So much for evolution's iconic lynchpin.

XI

Tree Roots

The Plant Kingdom's Discontinuity

"Poems are made by fools like me,

But only God can make a tree." [1]

Joyce Kilmer

© Mike Norton

Scraggly Bristlecone pines, stalwarts of the Plant Kingdom, thrive high in the Sierras with rings suggesting an age reaching more than 4,000+ years ago.

Evolution's imaginary "Tree of Life,"

neglected identifying a branch for 80,000 species of trees.

Even though Charles Darwin investigated the Plant Kingdom, his conjectured transitional chain of organic life seems to have given primary attention to the Animal Kingdom instead. While media hype touts dinosaurs-to-birds and molecule-to-man scenarios, evolution's grand scheme gives short shrift to the origin of grass, bushes, vines, and flowers.

Peaches, strawberries, roses, lettuce, walnuts, cacti, seaweed, pineapples, and Kansas Tall Corn all take root and grow under the same umbrella classification that sustains animal life.

While appreciative of the delicious flavors of berries and fruits and the delicate fragrance of flowers, Sgt. Alfred Joyce Kilmer, killed in World War I, left his poetic legacy, reminding humankind: "There is nothing lovely as a tree."

The 80,000 known species of trees collectively qualify as kings of the Plant Kingdom. From those scraggly bristlecone pines, clinging to centuries of life on craggy Sierra slopes, to the giant sequoias penetrating 300 feet into California skies, trees dominate earth's landscape.

As for their enhancing human and animal life, think on these things: shade on a hot day; shelter from a cold wind; wood to build a house; fire in the fireplace; a secure home for birds to perch and build nests; magnolia and cherry blossoms; fruit like apricots, figs, pears, plums, pineapples, grapes and mangoes; walnuts, almonds, pecans; raw material for coal; nectar for bees and a storage vault for honey.

The list goes on but you get the picture: Trees rule the Plant Kingdom.

For want of a valid scientific explanation, the chance hypothesis buys into the abstraction that the first living cell managed to create itself from non-living matter--by accident no less. Darwin takes another leap of faith, swallowing the indigestible thought that all plant and animal life evolved from the same make-believe, common ancestor.

"Analogy would lead me…to the belief that all animals and plants have descended from some one prototype…I should infer from analogy that probably all the organic beings which have ever lived on this earth descended from some one primordial form, into which life was first breathed." [2]

Let's get this straight--the most oft-quoted proponent of the chance hypothesis in the two most recent centuries pledged allegiance to the myth that trees and humans share descent from "one primordial form" ancestor.

"Modification through variation and natural selection" explains nothing.

Charles Darwin spent long hours with his botanist son researching plants, but his 1859 *Origin of Species*, featuring "descent with modification," neglects to clarify the genetic junction where strawberry plants and giant sequoias deviated from hummingbirds and elephants after having split from that original, self-created "primordial form" and redesigned themselves by "variation and natural selection."

So where's the genetic trail confirming a redwood tree and a monarch butterfly share a common ancestor?

Where's the fossil intermediate? And what did he, she, it, look like?

If Darwin's rash conjecture is to be believed, just where are the mega-millions of transitional life forms that might lend credibility to the thesis that trees and animals came from that "one primordial form?" Even open-ended deep time years of "slight successive variations" appear hopelessly inadequate to account for this level of genetic discontinuity.

Verifiable scientific evidence that the Animal and Plant Kingdoms both evolved from that identical first life seems lame and elusive.

Or were these two discontinuous living formats designed and introduced during creation week, mutually co-dependent components of a vibrant, global ecosystem? Could animals survive without grass, berries, fruit, vegetables, nuts, flowers, shrubs and trees? Could either living kingdom thrive and survive without the simultaneous presence of the other?

The oldest known life forms living today belong to the Plant Kingdom. The bristlecone pine *Methuselah* is reputed to have weathered more than 4,000 winters at the upper edge of the Sierra tree line.

© Mike Norton

An up close look at the exposed grain of s Bristlecone Pine.

One evolutionist credits land plants as *Methusaleh's* ancestors. "The first land plants…appeared 460 million years ago and trees evolved from those early herbaceous land plants…and have dominated terrestrial ecosystems for 370 million years…" [3]

Despite high-sounding phraseology, Henry Gee's cautionary comments come to mind: to take "...a line of fossils and claim that they represent a lineage" is comparable to an "assertion that carries the same validity as a bedtime story." [4]

More than 150 years after the publication of *Origin,* tree evolution remains anything but an exact science.

Trees "are critical to the biosphere...sheltering organisms, absorbing carbon dioxide, preventing erosion, producing oxygen, regulating climate, cycling and distributing crucial nutrients, providing raw materials, and cleaning the air." [5] Add to all this elegant beauty, food for humans, comforting shade, refuge for birds and animals, replenishment of soil with nutrients, as well as having a critical climate impact.

© Robert Paul VanBeets

Rather than transitional fossils, bits and pieces of what once had been living trees litter Arizona's high desert landscape in a colorful Petrified Forest.

While discontinuity separates plants and animals, wide genetic gaps divide trees from plants. The Plant Kingdom consists of 300,000 known species, of which 250,000 consist of flowering plants called angiosperms. This dominant category includes "roses, tomatoes, rhododendrons, the various grasses, and the flower trees such as sassafras, oak, palm, and apple." [6]

"Darwin's fascination with plants, which continued for almost 40 years after his first sketches of The Origin, appears to have begun with an early interest in pollination." [7]

No stranger to the Plant Kingdom and confronted with a genetic "mystery," he struggled to find a niche for trees and shrubs in his grand scheme.

Beyond inability to resolve the "far higher problem of the essence or origin of life," the "sudden" appearance "of the higher plants" in the fossil record mystified Darwin. Figuratively throwing up his hands, he acknowledged that "nothing is more extraordinary in the history of the Vegetable Kingdom…than the apparently very sudden or abrupt development of the higher plants." [8]

In an 1879 letter to Joseph Hooker, Darwin said that his plant research led him to confess perplexity that transitional ancestry leading to flowering plants, as required by his theory, seemed missing from the fossil record. "The rapid development, as far as we can judge, of all the higher plants within recent geological times is an abominable mystery." [9]

Darwin's attempt to design a "tree of life" did not begin to solve the "mystery." Some modern skeptics trash the whole idea.

"For a long time the Holy Grail was to build a tree of life. We have no evidence at all that the tree of life is a reality…in the past couple of years people have begun to free their minds…its time to move on." [10]

As with the fossil records, thin evidence of "organic chains" in the Animal Kingdom shows that discontinuity reigns supreme in the Plant Kingdom. Verifiable evidence of transitional life forms, ancestral to the Plant Kingdom, range from the dubious to the non-existent.

"...The fossil record does not support an evolutionary origin for trees from non-plant forms.... The first trees existing in the fossil record were clearly trees. Furthermore, an enormous gap exists between trees and all other plant forms." [11]

Discontinuity rather than a continuous chain of evolving organic life forms dominate fossil records of both plant and animal kingdoms. So where's the evidence plants evolved side-by-side with animals while sharing common ancestry with "one primordial form?"

© Alex Edmonds

Where did that Giant sequoia, towering a football field length piercing the sky, inherit its 3,500 year-old roots?

A mid-twentieth century University of Michigan Curator of Fossil Plants recorded his personal lament.

"It has long been hoped that extinct plants will ultimately reveal some of the stages through which existing groups have passed during the course of their development, but it must feely be admitted that his aspiration has been fulfilled to a very slight extent, even though paleobotanical research has been in progress for more than one hundred years." [12]

Late in the 20th century, a Cambridge University Botanist said that "to the unprejudiced, the fossil record of plants favours [sic] special creation." [13]

Despite attempts to theorize the origin of trees by a random chance process, "...continuity of trees arising from single cell precursors through evolutionary process has not been supported by the fossil evidence." [14]

Wollemi pine trees, conventionally dated at 150 million years ago, were long considered extinct. Yet, contrary to Darwin's dire prediction that "not one living species will transmit its unaltered likeness to a distant futurity," hardy Wollemi pines have been discovered, thriving in an obscure Australian grove, virtually unchanged from its ancient ancestry. [15]

Historic Old Colorado, John Herbert Johns unpublished portfolio

Remnants of trees, embedded deep in coal seams since ancient times, include a diverse roster of sturdy plant kingdom reps: sassafras, laurel, poplar, willow, maple, beech, birch and elm. Descendants of these familiar names survive virtually unchanged today, oblivious to Darwin's prediction of their certain demise.

Hickory, walnut, magnolia, and gingko trees, along with grape vines, and water lilies boast ancient fossil ancestries. essentially matching twenty-first century's distinctive counterparts.

No transitional intermediate from that alleged "primordial form" leading to trees has been certifiably identified. Nor have any new and distinctly different life forms evolved from Wollemi pines, Bristlecone pines, or Giant Sequoias in the last 4,000 years of recorded history.

Does deep time add anything to the stasis equation? Or is evolution no longer doing its thing, as if it ever squared with reality?

As to flowers such as crocus, roses, orchids, and the purple blossoms that spring from wisteria vines--can the chance hypothesis identify the common ancestor prior to the alleged "descent with modification?" For that matter, what chain of organic life forms evolved a cactus?

Then there are those food-producing plants that sustain animal life!

Potatoes that grow in the ground; walnuts, apples, peaches and apples that decorate trees; grains like wheat and corn; and vines that produce tomatoes, grapes and raspberries.

Darwin's "abominable mystery" remains unresolved!

Without evidence of tree "descent by modification," is it reasonable to conclude plant evolution ever occurred? If not, is then rational to claim the chance hypothesis applies to the Animal Kingdom, even in a diminished form?

Again, Joyce Kilmer's poetic phraseology resonates with reality: "Only God can make a tree!" [1]

XII

The Laughing Mouse

Irreducible Complexity

"…Natural selection can only choose systems that are already working…
if a biological system cannot be produced gradually
it would have to arise as an integrated unit, in one fell swoop,
for natural selection to have anything to act on." [1]

Michael J. Behe

© Jemini Joseph

The Paris Eiffel Tower, an icon of "irreducible complexity."

Irreducibly complex mechanisms, living or man-made, require a system of co-dependent components in order to function.

Today's world turns on personal computers. Take away the hard drive and the irreducibly complex machine becomes useless junk. Without a keyboard or the electronic mouse, the machine would not function.

Like man-made computers, animal and plant life forms could not exist without all DNA components in place. Biochemist Michael J. Behe coined landmark phraseology when he suggested molecular systems in living organisms are "irreducibly complex." He illustrates the world of molecular machines by citing the mousetrap as a prime example. A mousetrap is as simple as any human designed mechanism. When all its components function in unison, the trap destroys mice. Remove any single part, and the trap morphs into harmless fragments of matter.

The simple but "irreducibly complex" mousetrap consists of five basic parts anchored by four staples. It requires the complete, designed package of parts for minimal function. A spring too weak, a trigger too short, or a missing staple and the mouse might steal the cheese, survive unscathed and scoot away safely with the last laugh.

One skeptic suggests the trap could do without the wooden base by stapling the trap's metal parts to the floor. The argument fails, for the floor itself becomes an essential part of the mechanism.

A fully functioning mousetrap is the product of human intelligence. No one argues that inanimate machines design and build themselves without input from an intelligent source.

Design and construction of living molecular mechanisms also requires intelligent input. Living mechanisms display designs infinitely more complex than objects designed by human intelligence. Assembling operational living systems requires insight beyond nature's whims.

"Canyons separating everyday life forms have their counterparts in the canyons that separate biological systems on a microscopic scale. Like a fractal pattern in mathematics ...unbridgeable chasms occur even at the tiniest level of life." [2]

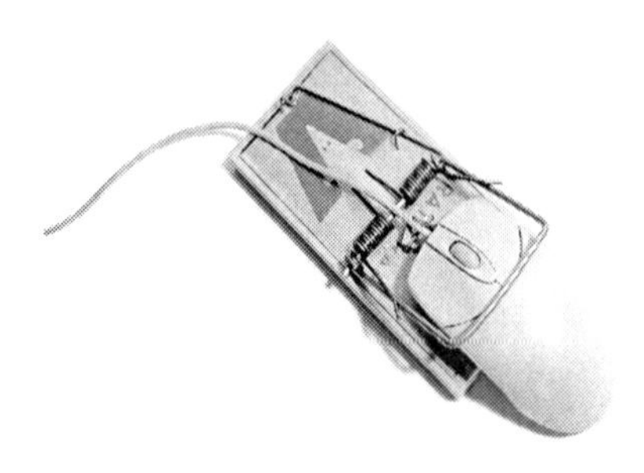

© Stephen Coburn

Whether whales transitioning from bears, or birds from dinosaurs, part-way-there transitional form couldn't survive a gradualist genetic journey.

A prime example of *irreducible complexity* in the living world is the bacterial flagellum, a molecular machine which functions with 40 separate components, 30 of which are unique.

The flagellum's microscopic propeller cranks at an amazing 100,000 RPM (Rotations oer Minute) and is capable of abruptly stopping and reversing on a quarter turn. If any single component is missing, the entire organism breaks down. Time is of the essence. It's an all-or-nothing scenario; each component must be in place simultaneously for the living mechanism to function. [3]

"Irreducible complexity" is the joker in evolution's card deck. Organs of living systems cannot evade this cardinal principal. Human eyes could not see without optic nerves and could not survive without the built-in luxury of being bathed by the lachrymal gland's soothing liquid.

Behe cites the chemical apparatus marshaled by the half-inch bombardier beetle as an irreducibly complex system in action. When threatened by an enemy it can release a scalding hot liquid as a defense.

"The components of the system are (1) hydrogen peroxide and hydroquinone, which are produced by the secretory lobes; (2) the enzyme catalysts, which are made by the ectodermal glands; (3) the collecting vesicle; (4) the sphincter muscle; (5) the explosion chamber; and (6) the outlet duct." [4]

Behe inquires: "What exactly are the stages of beetle evolution, in all their complex glory? Second, given these stages, how does Darwinism get us from one to the next?" [4]

"As the number of required parts increases, the difficulty of gradually putting the system together skyrockets, and the likelihood of indirect scenarios plummets.

"Darwin looks more and more forlorn." [5]

***Irreducibly complex* blood powers animal life.**

Custom-designed formulas match the needs of each species.

One-of-four blood types flow through human arteries and veins. Egyptian mummies share identical bloodlines with modern *Homo sapiens*, unchanged after thousands of years. Transfusing blood from ape to man can kill; extreme blood loss can kill; blood too thin to clot can kill; and blood that clots at the wrong time and place can kill.

Behe emphasizes the magical qualities of clotting.

"If blood congeals at the wrong time or place…the clot may block circulation as it does in heart attacks and strokes…A clot has to stop bleeding all along the length of the cut, sealing it completely. Yet blood clotting must be confined to the cut or the entire blood system of the animal might solidify, killing it." [6]

Properly functioning blood clotting in the human circulatory system that can save life involves no fewer than twenty steps.

"The blood coagulation cascade" utilizes "proteins in promoting clot formation" and proteins "involved in the prevention, localization, or removal of blood clots." [7]

"None of the cascade proteins are used for anything but controlling the formation of a blood clot. Yet in the absence of any one of the components, blood does not clot, and the system fails…Not only is the entire blood-clotting system irreducibly complex, but so is each step in the pathway." [8]

"Remember a mousetrap spring might in some way resemble a clock spring, and a crowbar might resemble a mousetrap hammer, but the similarities say nothing about how a mousetrap is produced.

"In order to claim that a species developed gradually by a Darwinian mechanism a person must show that the function of the system could 'have been formed by numerous successive, slight modifications.'" [9]

The statistics say, "No way!"

Blood clotting could not evolve "by numerous successive, slight modifications."

Tissue Plasminogen Activator binds to several substances including fibrin. TPA "has four different types of domains…the odds of getting those four domains together is 30,000 to the fourth power.

"Now if the Irish Sweepstakes had odds of winning of one-tenth to the eighteenth power, and if a million people played the lottery each year, it would take an average of about a thousand billion years before anyone (not just a particular person) won the lottery.

"A thousand billion years is roughly a hundred times the current estimate of the age of the universe." [10]

"We calculated the odds of getting TPA alone to be one-tenth to the eighteenth power; the odds of getting TPA and its activator together would be about one-tenth to the thirty-sixth power…

"Such an event would not be expected to happen even if the universe's ten-billion year life compressed into a single second and relived for every second for ten billion years…The fact is, no one on earth has the vaguest idea how the cascade coagulation came to be." [11]

Michael Denton, internationally recognized authority respecting molecular machinery, dismisses "gradual accumulation of random changes" as unsustainable in light of irreducible complexity.

"My fundamental problem with the theory is that there are so many highly complicated organs, systems and structures, from the nature of the lung of a bird, to the eye of the rock lobster, for which I cannot conceive of how these things have come about in terms of a gradual accumulation of random changes.

"It strikes me as being a flagrant denial of common sense to swallow that all these things were built up by accumulative small random changes. This is simply a nonsensical claim…a huge number of highly complex systems in nature cannot be plausibly accounted for in terms of a gradual build-up of small random mutations...

"Everybody knows the lung of the bird is unique in being a circulatory lung rather than a bellows lung…It doesn't require a great deal of profound knowledge of biology to see that an organ, which is so central to the physiology of any higher organism, its drastic modification in that way by a series of small events is almost inconceivable.

"This is something we can't throw under the carpet again because, basically, as Darwin said, if any organ can be shown to be incapable of being achieved gradually in little steps, his theory would be totally overthrown." [12]

The "break down" time that concerned Darwin knocks at the door.

Michael Denton's analysis takes Darwin at his word, challenging gradualism's imaginary make-believe as inadequate substitute for fact.

"If anyone was chasing a phantom or retreating from empiricism it was surely Darwin, who himself freely admitted that he had absolutely no hard empirical evidence that any of the major evolutionary transformations he proposed had ever actually occurred. It was Darwin himself, the evolutionist, who admitted in a letter to Asa Gray, that one's 'imagination must fill up the very wide blanks.'" [13]

Historic Old Colorado, John Herbert Johns unpublished portfolio

Colorado's intelligently designed Denver capitol is crowned with a thin layer of gold-leaf that could not shine without an irreducibly complex supporting dome in place.

Transiting animal life forms would be prime starvation candidates if left waiting for "numerous, successive, slight modifications" to evolve jawbone dentition essential to support a new and different organism.

Donald R. Moeller, a science researcher blessed with professional credentials as both physician and dentist, shoots down any possibility that unique dentition sizes and shapes result from mutations.

Moeller's logic devastates Neo-Darwinism's core dogma's reliance on "numerous, successive, slight modifications." Modern craniofacial/maxillofacial genetics demonstrates that pleiotropy is present in approximately one hundred genetic disorders.

"Thus the simplistic idea of genetic mutations being able to cause only incremental small useful changes in the occlusion and/or jaw relationships is not supported by current research. There are no known mutations affecting single tooth morphology or single tooth enamel microstructure.

"Multiple examples of malposed teeth, cysts in jaws, retained deciduous teeth, maxillary-mandibular growth and size discoordination, losses of entire classes of teeth, variation in eruption height of the various classes of teeth, tooth size arch-discrepancy variation, animal size, and tooth size coordination, should be in the fossil record.

"This evidence is not seen...

"There is also no known genetic or developmental process to suggest a legitimate mechanism to support an evolutionary basis for the development of the precision exhibited by the dental apparatus." [14]

Darwin assembled a wish list of uncorroborated assertions, predictions, preconceptions, and assumptions in weaving together the whole cloth of accidental life without design or designer.

He "ransacked other spheres of practical research work for ideas...

"But his whole resulting scheme remains, to this day, foreign to scientifically established zoology, since actual changes of species by such means are still unknown." [15]

All living mechanisms function irreducibly complex. Molecular mechanism design requires intelligence beyond the quirks of nature's whims. Assembling complex systems requires oversight. It's all or nothing with the human eye. Sight is contingent upon every lens, blood vessel, fully operational; the same with ears, hearts, and lungs. There is no reason for an infinitely wise and all-powerful Creator to require millions of years to assemble a brain or any life system. Recognition of Supreme power accepts the plausibility of a creation event at the command of the Lord God Almighty in the blink of an eye.

The eye of an arthropod also astounds. Arthropod eyes "have always been complex---and there have always been arthropods...The bee's ability to convey the location of a food source to fellow workers via a sophisticated 'dance' is legendary." [16]

Trilobites "could see an undistorted image under water…with undistorted vision in all directions…to determine distance…while at the same time have the optimum sensor for motion detection." [17]

© allslutsky

Dragonflies hover and move in all directions carried by four irreducibly complex gossamer wings.

Park Rangers at Tennessee's Reelfoot Lake explain that a bird (like the red-tailed hawk, blessed with acute vision extending a mile in any direction) couldn't survive in the wild with a single eye. Gradualist eye evolution, beginning with tissue theoretically able to recognize light, certainly couldn't do the trick. Blind-man's-bluff is not a game birds of prey can play.

A hawk flapping half-formed wings while trying to hunt with eyes scarcely able to detect night from day couldn't feed or protect itself. A less "evolved" parent, even if miraculously alive, couldn't help its crippled offspring.

A screech owl, surviving a collision with a car, lost an eye in a Reelfoot Lake accident. Park rangers rode to the rescue, nursing the owl and all other injured wildlife back to health. When recovery is complete, healthy raptors are released to again patrol lakefront skies.

Whether an eye, wing, or foot, however long or short its "evolution," the part could never function in half-done format. Extinction intrudes long before expiration of evolution's mega-million years of deep time.

Irreducible complexity comes into play ecologically. Swallows swoop to mud puddles, scavenging nest material. Eagles build 3-foot wide nests in treetops. Arrayed in multi-colored hues and oversize beaks, parrots mimic human sounds.

Virtually nothing, either in the fossil record, molecular biology, or today's experience, confirms existence of partially or non-functioning intermediates linking these magnificent birds to reptile ancestors.

Plant and animal life systems function co-dependently.
Trees contribute food and attract rainfall; water grows green grass;
and pasture-grazing cattle munch the grass canopy.

Life did not require vast chunks of deep time to originate abruptly, fully formed and in perfect balance. Multiple plant and animal kinds, appearing simultaneously and functioning in concert, provided an irreducibly complex eco-system from the beginning.

Gradualism is out the window if tested against this all-or-nothing measure. A chaotic composite of non-functioning organic systems would throw ecological balance out-of-kilter, triggering extinction.

Peacocks spread their iridescent feathers with ear-piercing screams, notifying neighborhoods of intruders, while announcing their unique, irreducibly complex status.

Natural selection works contrary to gradualism by rejecting a non-functioning component part. Normally, any incomplete but irreducibly complex cell, organ, or system would be discarded by natural selection, action directly counter to evolution's stairway to oblivion.

Prolonged failure to function is a recipe for the destruction of most any living system trapped interminably within a series of intermediate states. A pitiful intermediate creature attempting to evolve a wing from a leg, left unable to walk or to fly, is destined to die.

© Ruth Page Johns

Since completion in 1937, San Francisco's irreducibly complex Golden Gate Bridge depends on every thread of its steel cables for survival.

When the delicate balance of a complex life is rendered dysfunctional—the life terminates. Conversely, a new and different life format can't evolve from an organism missing critical functions (e. g., the arthropod's vision). Nature bursts with dynamic change, but irreducible complexity prevents activation of the natural selection anchoring the chance hypothesis. *Half-formed hearts and lungs are no more practical than half-built automobiles.*

Thanks to genes present from the beginning in every genome, life kinds adjust to preserve survival while accessing nature's smorgasbord of colors, shapes and sizes. Adjustment capability doesn't result from mutations but comes from the genetic code already inherent in every genome.

Darwin contemporary, Sir William Dawson, warned that when evolution was scrutinized, "the whole fabric would melt away like a vision. This evolutionist doctrine is itself one of the strangest phenomena of humanity...a system destitute of any shadow of proof, and supported merely by vague analogies and figures of speech...now no one pretends that they rest on facts actually observed...

"...Evolution, as an hypothesis, has no basis in experience or in scientific fact, and that its imagined series of transmutations has breaks which cannot be filled." [18]

Irreducible complexity underscores an all-or-nothing reality!

Sir William's candid assessment proved on the money. The century since Darwin cast his vote for molecule-to-man by accident has not been kind to the chance hypothesis.

Biologist Søren Løvtrup scoffed at evolution's unproven scenario.

"I believe," he said, "that one day the Darwinian myth will be ranked the greatest deceit in the history of science." [19]

XIII

An Inconvenient Truth

Extrapolation is Not Evolution

"...The fact remains that there exists to this day not a shred of bona fide evidence in support of the thesis that macro evolutionary transformations have ever occurred." [1]

Wolfgang Smith

Historic Old Colorado, John Herbert Johns unpublished portfolio

Kaleidoscopic change in the natural world infuses rhythmic balance in a dynamic ecosystem.

The sun and the moon deliver alternating shades of light and dark. Up or down shifts in temperature transforms water to steam or to ice. Seasonal climate changes trigger cycles where green leaves sprout from bare tree limbs en route to flashing a riot of colors each autumn.

Predictability is nature's common denominator, and it drives a fluctuating environment. Evolutionary theory contradicts this reality by substituting chaotic accident as the norm.

Lacking evidence, propagators of the chance hypothesis resort to extrapolating change from within a genome as proof that once corrupted by mutation, an organism can modify its identity and make a "biologic transit stop" en route to some radically different body plan.

Extrapolation doesn't change reality. Finches may modify beak sizes and shapes but they remain finches.

No matter how many times a pair of dice is tossed, the resulting number of dots will never exceed twelve. Even if repeated daily over 4.6 billion years of deep time, the odds of throwing more than a twelve-dot combination remain zero.

It would be nonsensical to extrapolate from the actual range potential of two to twelve dots as evidence that a mix of thirteen or higher is possible, that is, if given enough time. Citing a genome's known ability to rearrange its gene mix in order to survive as evidence confirming evolution is equally absurd.

Inherent in the genetic code of each genome is the capacity to adjust to ensure survival. But without adding new genetic information, intra-genomic change does nothing to produce radically different kinds of organic life. For instance, hundreds of generations of mutated fruit flies remain fruit flies. Bacteria change but never evolve into a life form other than bacteria.

Galapagos finches remain finches, never evolving into eagles, bears or whales. Changes accessing genetic information already present within the genome's gene pool do not correlate with evolution's microbe-to-man idea. Finches, bacteria and moths may exhibit variations resulting from the reshuffling of genes already present in their DNA information pools but descendants remain finches, bacteria and moths.

Bottom line reality: Intra-genomic change is not Darwin's evolution. Do any of these statements make scientific sense?

The size and shape of finch beaks may modify over time.

Bacteria can develop immunity to antibiotics.

Mutations can alter the number of fruit fly wings and legs.

A "yes" is the correct answer to all three statements. "Yes" answers don't make you an evolutionist. Creationists also answer "yes."

The kaleidoscope of life pulsates in synchronized motion. Each genomic life kind displays the capacity to adjust its size, shape, color, or chemical tolerance, expressing diversity while enhancing survivability.

This is not the quantitative and qualitative change evolution predicts.

So what's the big deal?

Do you believe a fish is the ancestor of your own family?

A "yes" to that one locks you inside the inner sanctum of evolution's pantheon of the make-believe. Extrapolating the reality of a genome's versatility as proof that descendants of a "fish" or an "Old World monkey" could eventually evolve into a human is dead wrong, an intellectual scam.

Molecule-to-man? Hardly. With transitional links still missing from fossil clues, living history remains evolution's last best chance to find exoneration. No matter how it's sliced and diced, several thousand years of recent history don't demonstrate evolution in action, either gradually or by giant leaps.

Gerald A. Kerkut recognized and addressed extrapolation with a rhetorical question and a definitive answer.

"Why can't one extrapolate and say this has in effect led to the changes we have seen right from the Viruses to the Mammals?

"Of course one can say that the small observable changes in modern species may be the sort of thing that lead to all the major changes, but what right have we to make such an extrapolation?" [2]

Yawning, genetic chasms separating distinct kinds of organic lives have never been bridged by evolution's natural selection, either in ancient rocks or in the living world. Mutations add nothing to the flawed equation. Look at *E. coli* bacteria. Modifications have been observed in 44,000 *E. coli* bacteria generations in the course of a twenty-year Michigan State University lab experiment launched in 1988. Limited initially to a glucose-only diet, descendant *E. coli* managed to adjust and to pull a molecule of citrate through their membrane and "thrive on it as their sole source of carbon." [3]

The day after news of the *E. Coli* lab experiment was posted in June, 2008, one skeptical blogger asked, 'Yeah, but its still a bacteria. Let me know when a bacteria [bacterium] evolves into a dog or a cat." [3]

Another doubter signed-in the following day, writing: "Are you aware of any long-running experiments like this where the initial bacterium has accumulated sufficient mutations that in the end it would be classified as a different type (genus or something higher) of bacterium from what it started out?" [3]

Good question. The answer is that it has never been observed.

An earlier (2002) news journal report described evolution as a "fundamental fact of biology," citing laboratory induced *E. coli* modifications as evidence. The report referenced laboratory findings of variations in twelve populations of *E. coli* bacteria that reproduced "every 3.5 hours or so."

After thousands of generations, once genetically identical populations had each "adapted in its own way to the conditions in its test-tube home." [4]

End of story?

E. coli populations remain *E. coli* because of built-in genetic information enabling heroic feats of adjustment. The laboratory findings point to the conclusion that *E coli* parents produce nothing but *E coli* offspring.

This *E. coli* experiment exposes an inconvenient truth: Every organism carries pre-loaded, genetic reserves enabling it to modify in order to survive. The genome's modification ability is certifiable fact but doesn't begin to verify Darwinian evolution.

Thousands of generations into the experiment, those pesky *E. coli* faithfully perpetuate their unique, *E. coli* persona, *ad infinitum.* Mutations may upset the genome, but lack of new genetic information bars radical, evolution-style change. Forty-four thousand generations of *E.coli* didn't do evolution's bidding in twenty years. Stasis intrudes, trumping evolution.

Do the odds improve after twenty thousand years? Or twenty million years? How about four billion years?

Is deep time the magic elixir that renders the impossible, possible?

The fact is that, 44,000 generations after the fact, descendants continue reproducing thousands of generations of unmistakable *E. coli.* Survival of modified *E. coli* represents the opposite of hocus-pocus transition to some radically different organism.

Genetic card decks may be reshuffled and genes may be lost or damaged, but without addition of new genetic information, a genome will not "evolve" into a radically new and different life format.

The ability of bacteria to shuffle genes to build descendants immune to antibiotics contributes nothing to Darwin's grand scheme. Bacteria can mutate and multiply to their heart's content, but they remain stuck in their bacteria mode.

The same goes for descendant populations of insect pests building resistance to pesticides--hundreds of generations down the line, the population remains an insect pest population.

The reality of this form of observable change is scientifically on the money and encompasses "bacterial antibiotic resistance, insect pesticide resistance, industrial melanism [peppered moth], sickle-cell anemia, and increased fitness in irradiated populations of Drosophila." [5] Thousands of years of living history reveal no persuasive evidence demonstrating evolution in action.

Genetic exuberance thrives, activating the adjustments essential to the survivability of the genome's unique identity. Every prototype life carries a staggering stash of genetic information, in place from the start. A genome carries the capacity to reach into its genetic reserves and switch its genes on or off to survive.

Biologist Jonathan Wells explains:

"...Mutations and natural selection are significant factors at the molecular level, especially in rendering bacteria resistant to antibiotics, or insects and other pests resistant to pesticides...Like antibiotic resistance, most insecticide resistance is due to inactivating enzymes," or "spontaneous mutations...Raw materials for large-scale evolution must be able to contribute to fundamental changes in an organism's shape or structure." [6]

Despite Jonathan Wells debunking the English peppered moth as an icon of evolution in 2000, this fake "proof" of evolution continues to be touted in some biology classrooms.

Ecologists at Alabama's Auburn University studied two populations of house finches that had recently moved into new habitats. One group of birds moved from New York to Alabama, the other from California to Montana. Each group rapidly adapted to its new climate and after thirty years the two populations showed differences in appearance and behavior.

In Alabama, males grew faster than females did and displayed wider bills and longer tails. In Montana females grew faster and were larger. An increased survival rate for offspring occurred in both locations. [7]

Despite the modifications, these house finches remained house finches.

In another experiment, scientists tested the limits of the fruit fly's ability to adapt to an increasingly dry environment. Starting with the most "desiccation resistant...they subjected the insects to very dry conditions until 90 percent had died, and then they bred the survivors." [8]

The remaining hardier ten percent were bred further for fifty more generations. The researchers expected to produce "even more dryness-tolerant flies. But what they got were flies basically no different from the ones straight out of the rainforest." [8]

Hundreds of generations of ubiquitous fruit flies, subjected to laboratory-induced mutations, may add or subtract wings and legs but continue producing fruit flies, stuck with their fruit fly pedigrees.

An evolutionary biologist looked at corn in Mexico showing "no resemblance to modern-day corn" and concluded, triumphantly, "If that's not evolution in action, I don't know what is." [9]

In his exuberance, the biologist ignored the obvious: Corn remains corn.

Science magazine touted: "Evolution in Action" as 2005's "Breakthrough of the Year." Declaring evolution dogma to be "the foundation of all biology," the assessment boasted "...every discovery in biology and medicine rests on it, in much the same way that all terrestrial vertebrates can trace their ancestry back to the first bold fishes to explore land." [10]

Attempting to fortify the case, the authors referenced the ability of the marine stickleback fish "...to adapt rapidly to a new environment." Warblers, corn borers, butterflies, crickets, cichlids, and, of course, the forever-favorite fruit fly were also touted as prime "breakthrough" examples.

The genome of the marine stickleback fish, like the genome of Mexican corn, enhances its survivability by accessing its own genetic reserves. Regardless of this obvious change, the surviving sticklebacks spend the remainder of their aquatic lives as marine, stickleback fish.

The same with the versatile genomes of warblers, corn borers, butterflies, crickets, and cichlids: The lineup of usual suspects, all candidates for coronation as "proof" of evolution, pop up regularly in science literature.

The public has endured repetitious media pronouncements and waded through biology textbooks citing population swings of gray and white peppered moths as "evidence" that evolution is real. But peppered moths still come in multi-hued gray and white colors, continuing as peppered moths, never parenting a completely new and different organism. The observed modifications do nothing to corroborate Darwin's unfulfilled prediction that "not one living species will transmit its unaltered likeness to a distant futurity."

Knee-jerk insistence that the expression of genetic reserves confirms evolution simply misleads the public and corrupts science. The potential flexibility, inherent in every plant and animal, is limited to the expression of its own pre-existing genetic code, and not a transition to another organic kind.

Dazzling diversity and dynamic versatility, for sure; but never a biologic transit up-the-down staircase to different families, orders, and classes--all topped with a giant leap to a different phylum. Hybrids, as per Gregor Mendel, mix genetic information assuring diversity.

The canny Clarence Darrow laced the 1925 Scopes trial written record with a semantic ruse, arguing hybrids confirm evolution, far less than the whole truth and nothing but.

Johns Hopkins University's Maynard M. Metcalf rode the coattails of hybridization citing a plant breeder's skill at modifying vegetables or flowers as proof of "evolution...occurring today...under man's control." [11]

Not the random chance radical change in nature, independent of an intelligent act required by evolution theory, although thought provoking enough to make the court record.

"Not only has evolution occurred; it is occurring today and occurring even under man's control. If one wishes a new vegetable or a new flower it is within limits, true that he can order it from the plant breeder and in a few years he will produce it...This is evolution of just the sort that has always occurred." [11]

Just because a high jumper breaks Olympic records doesn't mean his descendants might someday hurdle the moon, if given mega-time and millions of attempts. Extrapolated intellectual apparitions vanish as certainly as a drifting fog melts when kissed by the morning sun.

Despite blatant misinterpretation in the Scopes trial record, Gregor Mendel and Luther Burbank's hybridization represents true science. The size of tomatoes may be increased and the taste of apples improved, but this capacity for genetic change has nothing to do with evolution.

The much coveted Honey Bell tangelo is a hybrid cross from the Dancy tangerine and the Duncan grapefruit; is seedless, shaped like a bell, fiery-gold, and sweeter than an orange. Genes mixed from related organisms are capable of producing hybrids in nature, and without human intervention of a human mind.

The London Times announced the début of a hybrid weed, erroneously tagging the weed with the "evolution" label. The report told of a Scotland-based plant evolutionary biologist, who had checked the DNA of a weed found in York, and identified the plant as a natural hybrid between the common Groundsel and the Oxford Ragwort. Since the weed produces fertile offspring, and does not breed with parent species, the *Times* characterized the process as "evolution in action." The news daily crowed, "Charles Darwin was right and the creationists are wrong...the first new species to have evolved naturally in Britain the past 50 years." [12]

The enthusiastic report implying "Eureka," overlooked the obvious: The hybrid continued its lowly weed status. The weed that is still a weed didn't move its classification up the ladder to family, order, or class.

Despite extravagant headlines, extrapolation from hybrid reality can't verify scientific impossibility.

DNA gene information is not vulnerable to accidental corruption of genetic codes, leading organisms recklessly to some unpredictable "biologic transit stop." Wheat, corn, chickens, strawberries, and dogs can be bred to size, shape, and color by shuffling the gene cards. But genetic codes define limits to change unless something like gene splicing (guided by intelligence) introduces new information into the genome. Modified products continue as selected strains of wheat, corn, chickens, strawberries, and dogs.

"...The degree of change that can be experimentally induced in a wide variety of organisms, from bacteria to mammals, even under the most intensive selection pressures, is always limited by a distinct barrier beyond which further change is impossible." [13]

Luther Burbank made the case alluding to what he defined as "Reversion to the Average." "...It is hopeless to try to get a plum the size of a small pea, or one as big as a grapefruit ...there are limits to the development possible, and these limits follow a law." [14]

Plant breeders have been able to increase sugar content in sugar beets from 6% to 17%. [15]

Apple growers managed to convert the Hawkeye, "a round, blushed yellow apple of surpassing sweetness," discovered in 1880 in a Madison County, Iowa orchard, into Washington's once popular Red Delicious, with a five points base. "Breeders and nurseries patented and propagated the most rubied mutations" altering the color, shape, flavor, and juiciness. [16]

Genetically modified crops have become a fact of life in today's agriculture. New genetic information has been introduced to enhance production and profits of corn, sugar beet, cotton and soybeans.

"All dogs from the smallest Chihuahua to the biggest Great Dane emerge from the same basic set of genes. At the DNA level, two randomly chosen dogs differ by only about as much as two randomly chosen people do, yet the variation in appearance, size and behavior in dogs is just mind-boggling." [17]

All dog breeds, recognized and registered by the American Kennel Club, belong to the same *Canis familiaris* species, fully capable of interbreeding to produce lovable "mutts," treasured by their caretakers. By balancing the use of dominant and recessive genes, selected from components of the original gene pool, striking variations can result in size, appearance and temperament.

According to the Westminster Kennel Club, an estimated four hundred species of dogs claim descent from common canine ancestry. Courtesy of selective gene mixing, collies and poodles look different, but continue as dogs, man's favorite animal companion.

© pixshots

A dog…is a dog…is a dog. These guys may not look alike but they are hybrids from the identical species.

The marvelously diverse dog breeds answer to the same family heritage, never producing a feline or some representative of another animal class. This inherent genetic versatility does not make the cut as evolution. Intelligent human breeders skillfully play with genes of dogs, orchids, and pigeons in a continuing quest to produce unique colors, sizes, and shapes.

Imposing intelligent exercise in genetic design is not evolution. Hybrid variance in descendant generations of dogs didn't result from an accident in nature but from an intelligently directed breeding process that selects and develops desired traits from a gene pool in place from the beginning.

Dog breeding illustrates Mendel's law of genetics in action.

The complete *Canis familiaris* genetic code, deciphered in 2005, carries infinitely more diversity potential than components of a color wheel, a musical octave or the alphabet of a language. Building from a targeted portion of the master gene pool develops desired traits.

Selective access to any combination of the existing genetic code of the original parents assures variety without the addition of new genetic information or the crippling consequences of mutations. Variations radiate laterally within the species or sub-species, never climbing up the genomic staircase above and beyond the dog family.

Legendary English Foxhounds were bred to encompass the desired characteristics of the Greyhound, Bulldog, and Fox Terrier. Less stocky and a bit faster than its English ancestor, the American foxhound leads the way for Kentucky and Maryland red fox hunts.

"July," a hound prominent in American dog lore, was shipped from Howard County, Maryland to Georgia in 1858 where, after further crossing with other hounds, it sired the much-prized July-Maryland Foxhound strain of hunting dogs. After several centuries of breeding, Foxhounds, like all other recognized and unrecognized breeds, continue as components of the same descendant canines, never evolving into another living format.

No matter the genetic mix within the species, dogs never evolve into any radically different kind of life format other than the "*Dog*" family, "*Canis*" Genus, and species "*Familiaris.*"

Dogs (even the legendary "Heinz 57" breed) forge irresistible bonds with humans for life. Within days after loss of a beloved pet rescued from a shelter 14 years earlier, one dog lover penned a touching tribute to Kala.

"It may seem unusual to grieve the loss of a pet so deeply, but Kala was a truly exceptional animal. Even folks that were not dog lovers were amazed at her seeming human sensibilities. She was courteous and kind, a superb athlete, and a comedian.

"In my father's words she was, 'Just nearly perfect, aside from the fact that she does shed quite a bit.' "[18]

Recent scientific setbacks gnaw away at evolution's core fabric.

"In a year [2009] in which Darwin's disciples were celebrating the 200th anniversary of his birth...(1) the modern synthesis was dead, (2) Darwin's tree of life should be abandoned, (3) new "missing links" were a bust, (4) limits to Darwinism were demonstrated in the lab, (5) evolutionary icons like the peppered moths reverted back to their old colors, (6) the Cambrian Explosion lacks any plausible materialist explanation, and (7) an interdisciplinary revolution is occurring in biology that rejects the reductionist paradigm of Darwinian evolution."[19]

Darwin drew a blank trying to explain the origin of first life on earth. While his theory didn't attempt to explain life's origin, his reference to the "warm little pond"gave the nod to accident rather than design.

Molecular biology's electron microscope uncovered the magnificent complexity of the living cell with DNA's language of life, demolishing the primitive notion of the "simple" cell.

Darwin's fictional chain of organic life shows a fossil cupboard virtually bare. Gaps dominate. Discontinuity reigns. Evolutionist Ernst Mayr concedes, "All species are separated from each other by bridgeless gaps; intermediates between species are not observed."[20]

The Modern Synthetic Theory of Evolution dug deep attempting to salvage Darwinian thought by coupling *mutations* with *natural selection* as twin forces driving evolutionary change. This one-step-forward "solution" exposes a trail of two-step-back deficiencies.

(1) Mutations typically corrupt genetic information, never adding new genetic information and rarely upgrading the quality of the organism.

(2) Five-hundred-million years is not enough time to accumulate the billions of allegedly "beneficial" mutations essential for random chance to accidentally evolve an *irreducibly complex* human from a fish.

(3) Extrapolation of the real as proof of the never-was and never-will-be perpetuates unsubstantiated myth.

XIV

Playing the Time Card

Radiometric Dating

"Researchers have uncovered biological molecules like proteins, DNA, and pigments from rocks that are supposedly millions of years old...Many of these materials...will only survive thousands, not millions, of years." [1]

Brian Thomas

© Warren L. Johns

Deep time represents the mother lode of evolution theory. Without mega-millions of years, evolutionary conjecture washes away like fool's gold in a mountain stream.

More than 150 years after Charles Darwin challenged the Genesis account on the origin of life on Earth, his conjectures still defy validation.

Ever since he first floated his ideas publicly in 1859, scientific evidence has been unkind to his conjecture. By tests he himself set up, the case for evolution hangs in limbo. Given the shaky house of cards, evolution theorists look the other way and play the time card. All chips remaining from the diminished evidence pile are in play.

"Natural selection," purported to "advance by short and sure, though slow, steps," requires multi-millions of deep time years to work its transition myth up the ladder--from microbes to fish, to amphibians, to reptiles, to mammals, and ultimately to humans.

Aided and abetted by a complicit media, multi-millions of deep time years has emerged as a favored strategy to skirt the yawning gaps in evolution's unproven postulate. Time calibration defines an updated cutting-edge issue, rising to lynchpin status in pre-history analysis.

Geochronology is ranked an ideological touchstone. Nineteenth-century evolutionists conjectured 25 million years as time enough for random chance gradualism to work evolution's wonders. This number has been overwritten by today's guesstimates.

Debunked, dismantled and disarticulated, evolution advocates try to squeeze the concept through an academic back door. Open-ended deep time is an imperative to support Darwin's prediction that "all corporeal and mental endowments will tend to progress to perfection." [2]

The daily miracle of live human births flies in the face of evolution theory. Darwin estimated a time lapse of millions of years before that first living cell from some "warm little pond" would spawn complex *Homo sapiens.*

It takes only nine months for a single sperm cell to unite with a single egg cell to produce a living human baby. In several thousand years of recorded history, human reproduction has never shown the least hint of evolving a life at variance with the gene pool inherited from two human parents.

Even deep time may not be deep enough to salvage the legend of life's accidental origin and transitional development. The numbers game is anything but slam-dunk. Nor is it rocket science.

Whether measured in thousands or millions of years, time alone can't resuscitate a DOA concept infected with congenital defects. Screened by probability mathematics, the theory's *iffiness* collapses in a cloud of cosmic dust with or without deep time in play.

Even 4.5 billion years is not enough time to ratify medieval imaginings. Evolution's deep time placebo is a mirage.

Extremely improbable?

How about impossible!

As iconic gradualism crumbles, geochronology science has emerged to measure decay rates of inorganic elements. Ernest Rutherford introduced radiometric dating in 1905 as a methodology to determine the earth's age. Facing assaults on a tradition in jeopardy, die-hard Darwinists seized the new technology as a collateral challenge to the Genesis account of origins. The reasoning: "How could the Scripture narrative be true if radiometric dating produced multi-million year numbers?"

Although not slam-dunk, the answer resonates fact.

For example, no rational person would buy into the suggestion that a deceased family pet, buried on the "back forty," would share a birthdate with the ancient soil in the pasture. Similarly, the date of death of a fossil residue of a once living organism does not necessarily take on the geochronological age of the burial site's inorganic matter.

Think about it. Does it make scientific sense to assign fossil remains the same age as the surrounding burial site turf, composing the site of its discovery, just to accommodate evolution's geologic column?

Does the radioisotope age characteristic of the rock that enshrouds a fossil cemetery consistently represent the actual date of the death and burial event?

Since the answer is an obvious "no," evolutionary theory is deprived of an already fragile hope for salvaging serious semblance of scientific substance. Deprived of the radiometric dating crutch, evolution lacks any tool to confirm life on earth existed millions of years before the present.

While the authoritative phrase, *radiometric dating,* can intimidate, it can be plagued also by nagging discordance or by dancing with dates outside the accuracy range of a technology's limits.

Radiocarbon dating illustrates the challenge. The outer limit of C-14 dating is measured only in "thousands" rather than "millions" of years, a result not helpful to evolutionary theory.

"The most crucial dates in modern human evolution are unfortunately beyond the range of the radiocarbon method…" [3] C-14 dating of fossilized life forms determines maximum ages, but never "older than their C-14 content would allow." [4]

When a plant or animal dies, it no longer takes in C-14 but its C-12 remains constant. The comparative ratio of the two and the rate of decay is used conventionally as a reference tool for calculating the date of death. But determining the C-14 to C-12 ratio in order to calibrate antiquity of a fossilized life form requires an accurate starting date. If the actual time of day is 6 A.M., but a clock is set erroneously at 12 midnight, the clock will never show the correct time, however synchronously it ticks.

In the event the C-14 in the atmosphere is increasing faster than it is decaying, the C-14 to C-12 ratio would require readjustment and the C-14 time "clock" reset. Even if the ratio remains constant and stable over time, the outside limits of the findings would be measured in only thousands, not millions of years.

"Carbon-14 is used to date dead plants and animals because plants and animals incorporate C-14 into their bodies by eating, drinking, and breathing in an environment containing C-14...

"When the organism dies, however, it ceases to incorporate carbon into its body...Both C-14 and CC-14 to C-12 decreases slowly for thousands of years after the death of the organism." [4]

Coal seams, fossil residue from trees, and vegetation buried deep in the geologic column may not be as ancient as conventional radiometric dating techniques once suggested. Coal samples, deemed "millions-of-years" old by ^{14}C dating methods, should be ^{14}C dead, but they are not. In 1988, physicist Robert H. Brown reported on the unexpected presence of ^{14}C in coal seams previously dated in multi-million-year time zones.

"Infinite age' samples such as anthracite coal from deep mines in Carboniferous geologic formations (270-350 million years conventional age assignment) have yielded AMS C-14 ages in the 40,000-year range at laboratories in Europe, Canada, and the U.S.A." [5]

Similarly, "Paleozoic and Mesozoic coal and oil dates with the Accelerator Mass Spectrometer (AMS) method ...give maximum ages between 50,000 and 70,000 years. This indicates that they still have ^{14}C and seem to be younger than 70,000 years." [6]

Using the AMS method and calculating ^{14}C's half-life at 5730 years, John R. Baumgardner's 2003 findings recognized the raw measurement of the ^{14}C/^{12}C ratio improved from about "...1% of the modern value to about 0.001%, extending the theoretical range of sensitivity from about 40,000 years to about 90,000 years." [7] Baumgardner obtained ten test coal samples from the U.S. Department of Energy Coal Sample Bank: "...three coals from the Eocene part of the geological record, three from the Cretaceous, and four from the Pennsylvanian." [7]

The ten samples represented a 200-million-year time spread measured by conventional geological methods. The coal should be ^{14}C dead given the AMS limits of the 90,000-year measurement for traces of ^{14}C.

Instead of confirming absence of detectable levels of ^{14}C, the AMS tests disclosed "...remarkably similar values of 0.26 percent modern carbon (pmc) for Eocene, 0.21pmc for Cretaceous, and 0.23 pmc for Pennsylvanian ... little difference in ^{14}C level as a function of position in the geological record." [7]

Discrepancies plague traditional assumptions.

"...Penguins living in the Antarctic today have yielded 3000 year old carbon 14 ages ...Seals killed recently have ages of 1000 years..." [4]

"Living mollusk shells have been dated by the C-14 method at up to 2,300 years...a freshly killed seal at 1,300 years, and wood from a growing tree at 10,000 years." [8]

Fossil dating errors can be compounded when the assumed date of a fossil's death is based on the radiometric date of the burial site matter and then transposed to date a similar fossil at another global site believed to occupy a comparable place in the geologic column.

Assumption based extrapolation represents circular reasoning.

Uranium-Thorium-Lead, Rubidium-Strontium, and Potassium-Argon dating methods can generate discordant dates from identical test samples.

Gunter Faure warned of such *discordance* in isotope geology.

"Unquestionably, 'discordance' of mineral dates is more common than concordance...the mineral dates generally are not reliable indicators of the age of the rock.

"Although examples of nearly concordant U, Th-Pb dates can be found in the literature...in most cases U-and Th-bearing significance is questionable." [9]

Richard Leakey discovered what he believed to be a human skull in Kenya below rock "securely" dated at 2.6 million years ago. Radiometric dating of the KBS Tuff site ranged in erratic extremes from 0.52 million BP to 17.5 million BP. [10]

Was Leakey's burial site find correctly dated at 2.6 million years BP?

Was the fossil assigned an age based upon the assumed age of the burial site, or was the site assigned a date tied to the conjectured age of the fossil?

Is either assigned date accurate?

Or could this be extrapolated, circular reasoning reminiscent of the chicken/egg syndrome?

"Since formations we study today inherit radioisotope features from previous formations (erosion processes forming sediments; volcanic process forming igneous deposits), there is uncertainty as to how much of the daughter-product concentration in the formations we study accumulated during the geologic lifetime of those formations, and how much of the current daughter-product concentration was inherited from their source material.

"The well-defined starting point for radioisotope age determination does not assure a relationship between the radioisotope 'age' and the geological age (true time of existence) of a specimen." [11]

Discordant dating plagues Hawaiian Island time measurements. Using the potassium-argon method, ages ranging from 160 million to 2.96 billion years for lava flows that occurred in the year 1800[1] have been obtained. [12]

A cross-section of lava specimens taken from New Zealand's Mt. Ngauruhoe volcanic eruptions in 1949, 1954, and 1975 show potassium-argon dates ranging from a more recent 270,000 years BP to a distant 3,500,000 years in the past. [13]

When twenty-seven Brahma amphibolite samples of Grand Canyon basalt lava were collected from comparable sites and submitted to two "well-credentialed internationally-recognized, commercial laboratories for radioisotope analysis," the results indicated pronounced discordance.

"When the calculated error margins are taken into account the different radioisotope dating methods yield completely different 'ages' that cannot be reconciled---1240±84Ma (Rb-Sr), 1655±40 Ma (Sm-Nd), and 1833±53 Ma (Pb-Pb).

Historic Old Colorado, John Herbert Johns unpublished portfolio

Thanks to the early vintage car in the photo, the date of this Colorado snow scene can be approximated. Dating a young fossil sandwiched between layers of ancient rock is no slam-dunk.

"The K-Ar model 'ages' are so widely divergent from one another (ranging from 405.1±10 Ma to 2574.2±73 Ma), even from very closely spaced samples from the same outcrop of the same original lava flow, as to be useless for 'dating' any event...

"Irreconcilable disagreement within and between the methods is the norm, even at the outcrop scale. This is a devastating 'blow' to the long ages that are foundational to uniformitarian geology and evolutionary biology." [14]

Radiometric dating's discordance is not so shocking given tectonic dislocation of continents and radical magnetic pole shifts. Discordance lurks as an open secret, out-of-sync with time-certain events. Many deep time calibrations offer imprecise approximations that can't guarantee the absolute. Nor does ^{14}Carbon dating explain soft-tissue discovered in fossils imagined to be many millions of years old.

Conventional geochronology took a jolt in 2004 when respected paleontologist Mary H. Schweitzer reported the discovery of soft-tissue inside a fragment of a dinosaur femur unearthed in Montana. Implications sent rippling waves across time projections, real and imagined. [15]

"Mary Schweitzer found fresh T. Rex femurs in 1991 and 2000, and a hadrosaur femur with blood cells in 2009. She told Science in 1993, 'it was exactly like looking at a slice of modern bone.

"But, of course, I couldn't believe it. The bones, after all, are 65 million years old. How could blood cells survive that long?'" [16]

What about those alleged 65 million-year-old "highly fibrous… flexible…resilient" T. rex fossil fragments displaying "blood cells?"

"Blood cells" avoiding decay for 65 million years?

Billed as "unquestionably one of the most unexpected and important dinosaur discoveries of all time," the début of Leonardo, a mummified hadrosaur, raised academic eyebrows. More than a composite of "scattered collagen fibers," the long-extinct reptile displayed "skin pattern and stomach contents…discernable," with "whole tissues---in fact its whole body---still intact!" [17]

Supposedly having lived and died 77 million years ago, *Lenny* defies conventional time-frame expectations. With other recent discoveries, this find runs counter to multi-million-year scenarios for life-on-earth.

Collagen protein supposedly turns to dust within 30,000 years. But now its been reported that some collagen has been found as "fossil" material.

"DNA is particularly prone to decay, yet ancient fossil 'plants, bacteria, mammals, Neanderthals, and other archaic humans have had short aDNA sequences identified." [18]

Dinosaurs, the theory says, went extinct 65 or 66 million years before the present, leaving behind only fossil remnants. Then along comes *Dakota,* a hippo-sized, duck-billed hadrosaur unearthed by teenager Tyler Lawson on his family's North Dakota property in 1999. [19] Apparently *Dakota* didn't get the word that multi-millions of years of fossilization is more than enough time to erase all traces of organic matter.

Instead, Phil Manning's University of Manchester team "found that although the proteins that made up the hadrosaur's skin had degraded, the amino acid building blocks that once made up the proteins were still present. They believe that the dinosaur fell into a watery grave, with little oxygen present to speed along the decay process." [20]

The "watery grave" reference implies a collateral question: Could that hardrosaur have been a flood victim?

In the colorful phraseology of the astonished Dr. Manning, "You're looking at cell-like structures; you slice through this and you're looking at the cell structure of dinosaur skin. That is absolutely gobsmacking." [20]

There's more to the story. The "gobsmacking" picture conflicts with a sacred cow critical to chance hypothesis interpretations. Could organic material survive 66 million years after the actual date of death and burial? That's mega time of 660,000 centuries of 100-years each.

Then along comes Shixue Hu, of China's Chengdu Geological Center, reporting the discovery of a fossil cemetery of 20,000 fossils, many fully intact, buried fifty feet deep in a Luoping mountain in southwestern China.

Duly awarded a 250 million year age, the treasure trove of antiquity included stunning evidence out-of-kilter with traditional evolution. Rather than life forms radically different than today's fossil world, clams, oysters, snails, and the coelacanth testify to stasis.

But the stunner was the "soft tissues" found on some of the fossils!!! [21]

"Soft tissues" that escaped decay for 250 million years? Really?

Either dating methodology needs major-league fine-tuning or dinosaurs walked the planet more recently than many have imagined.

Circular reasoning projects bootstrap science, a conclusion based on conjecture. It may contribute to a fictitious paper trail, but it delivers nothing more than a guesstimate derived from assumption based extrapolation. Darwin's evolution dream remains nothing more nor less than what it has always been, a primitive, superstitious philosophy, scientifically unproven.

"Surprising as it may seem, the only real evidence for the geological succession of life, as represented by the timetable, is found in the mind of the geologist and on the paper upon which the chart is drawn. Nowhere in the earth is the complete succession of fossils found as they are portrayed in the chart…" [22]

The worst may be yet to come for devout evolutionists. What if a few thousand years in the past, the creation week of seven literal days, introduced a young life ecosystem to a much older blob of water-covered matter floating in cosmic darkness? Given that scenario, the radiometric age of earth's inanimate matter would prove irrelevant in determining the date of the creation of life on earth.

Under any scenario, evolution's time card is a joker.

XV

"Let there be light..."

Timelines

"I know this world is ruled by infinite intelligence. Everything that surrounds us - everything that exists - proves that there are infinite laws behind it. There can be no denying this fact.
It is mathematical in its precision." [1]
Thomas A. Edison

© Muharrenz

Ancient civilizations devised stone sun dials to calibrate time.

"And God said, 'Let there be light,' ...and there was light." [2]

Creation week marked the beginning of all life on earth. The first creation week miracle banished darkness.

It is no coincidence that on the first day of creation week the presence of God's Spirit appeared in a blaze of glory, introducing the light of the world that overwhelmed black darkness from the face of that blob of water-covered matter drifting in cosmic space.

The everlasting light, enveloped earth at the beginning of creation week, announcing the presence and the power of "the Spirit of God hovering over the waters." [3]

This day-one demonstration of Divine Authority, introducing creation week, confirmed God's presence and His pre-eminence as the source of universal light, energy, matter and life.

It wasn't until three days later that God created the sun and the intricate balance of the Solar System.

Had the sun been created on that first day rather than deferred until day four, humans might have been more inclined to bowing down to the shimmering rays of the servant sun rather than worshipping at the feet of the Master source of light and Creator of life.

Instead of honoring a fictitious sun god or a man-made inanimate object carved from wood or chiseled from stone, small-minded humans tended to drift from truth, overlooking the origin and essence of life. Eventually, blinded by intense rays from the solar inferno, some cultures chose to worship the seductive power of the sun while ignoring the fact that the celestial furnace was put in place and stoked perpetually by the Creator.

No inanimate object, stone idol or graven image deserves worship!

Instead of the hollow ritual of a "resounding gong or clanging cymbal," [4] God reserved a weekly day in time's march for mankind to rest from physical labor and to cherish the honor of being created in God's image. While no human walking today's earth witnessed earth's birth or the exact date of humankind's birthday, the weekly day of rest is a regular reminder that commemorates the event by honoring the Creator.

By lighting a darkness-saturated earth through the arrival of God's Spirit, three days before the sun existed, God has shown that He is the source of all light and life. With earth turning on its axis, there would be an evening and morning with the light emanating from God's Spirit.

Christ assured humans of access to the same power source that created the sun. "You are the light of the world…Let your light shine before men, that they may see your good deeds and praise your Father in heaven." [5]

So it wasn't until day four of creation week that "God said, 'Let there be lights in the expanse of the sky to separate the day from the night, and let them serve as signs to mark seasons and days and years.'" [6] This stellar event included opening skies to a star-filled heaven with a planetary system orbiting the sun.

More than dividing darkness from light, sun power sustains a living ecosystem, delivers invisible rays, controls climate, drives photosynthesis, filters water, cleans air, and provides earthlings with a reliable methodology for calibrating time.

Identification of a cosmic instant

requires measurement of light focusing on a moving target.

Earth-based chronology lacks the capacity to capture and identify a cosmic instant of universal time. But from within earth's limited perspective, calibration of time, motion, speed and distance relates to angles of light. Days, months, seasons, years and light-year distances correlate the here and "now" with history's "then."

While humans crave knowledge, attempts to pinpoint the precise moment of earth's birthday challenges finite thought. Curiosity propels minds to dig deeper to understand life's origin, purpose, and direction. This historic query opens a cavalcade of legitimate wonder.

Just how far back in ancient time zones did the Blue Planet and its miracle of life first début?

Insightful question. But this thirst for wisdom comes with a caveat: It runs the risk of exposing even the brightest human minds to gross fakery propagated by cunning prevaricators.

© Anderly Nekrasov

Attempts to measure unlimited stretches of uncharted past time strains comprehension. The enigma of pre-history extends beyond thought horizons while inspiring major modicums of faith. Universal time flows perpetually as an infinite flat line unless the Creator-God, the Designer of all time clocks, provides cohesion to the equation.

In the universal scheme of things, there is no such thing as time apart from a simultaneous "now" instant. Radiometric dating is not the cure all. Ultimately, the core issue relating to earth's age and life's origin is based on belief, or non-belief, in an infinite, all-knowing, all-powerful, Creator God.

The speed of dark can't be measured. Darkness is simply the absence of light and is useless in computing time. Tucked in a galaxy of precision, humans measure the time of day, relying on earth's turning on its axis and its orbiting the sun.

Readily identifiable *days* and *years* provide a measurement tailor-made for human comprehension. There would be no such thing as days, months, or years to measure earth time without the sun's light, the moon, and the "night clock" based on the precision of stars and constellations spinning around the North Star in a predictable east-to-west rotation. Earth's time technology is unique to this planet, without relevance beyond its niche.

Time doesn't "fly" the same way on other planets, much less in an infinite cosmos, without known boundaries, cluttered as it is with billions of spheres, all rotating in synchronized orbits. Since all Solar System planets spin in appreciably variant cycles, an instant on earth doesn't equate with time measurement on sister planets Venus, Mars or Saturn.

Mayans devised a calendar accurate to December 21, 2012. Long before access to computer technology, ancient minds "calculated the length of a year as 365.2420 days long," astonishingly close to modern astronomical time of 365.2422 days. [7]

God could have created all living things at His command, in the blink of an eye. Instead, to help humans recognize His power, He designed the sequential process in an easy to understand time environment.

The Genesis narrative depicts life created within the context of an irreducibly complex ecosystem built upon "formless and empty" matter when "darkness" covered "the surface of the deep." [8]

At the command of the Creator, a fully functioning ecosystem, complete with light, an oxygen- based atmosphere, dry land capable of supporting vegetation, the Plant Kingdom, and the Solar System—was put in place in the during the first four days of that creation week.

The Animal Kingdom, with birds and fish, followed on day five with land animals, and humans, the crowning achievement made in God's image on day six. Ever since day four, not only does sunshine chase away darkness every 24-hours, it paints the landscape with a kaleidoscope of dazzling colors guaranteed to brighten human psyches.

This daily show is but a hint of what that "lucky ol' sun" does to sustain and enhance plant and animal life. Earth's proximity to the sun impacts climate. Raindrops fall from sun induced evaporation. The world's food supply relies on the sun to do its thing.

The regular night/day sequence plays to human need for regular rest. Sleep comes naturally when darkness swallows the sun's rays.

Time calibration is based on earth-days and relative distances are tied to the speed of light. Measurement of time by days, and earth cycle orbiting the sun, provide a readily monitored time calculation.

A ray of earth-bound sunshine, traveling at 186,282 miles per second, takes 8 light minutes to reach Planet Earth. By comparison, some galaxies are estimated to be more than 13 billion light years distant. While intergalactic distance calibrations are subject to interpretation and error, the magnitude staggers minds.

This giant fireball delivers life-sustaining power to earth, fueled every day by what seems to be an unlimited tank of atomic energy. Moving through space in a path so precise that its location can be pinpointed hundreds of years in the past and predicted a thousand years into the future, the sun is so reliable in its path that its reliability is not subject to question.

Planet Earth seems designed to be an on-going adventure for human beings to be challenged by the phenomenal complexity of nature and the unfathomable "mystery of Godliness." [9]

The infinite, eternal God is just, loving, all-powerful, omniscient, and omnipresent. God pre-existed His creation.

Its axiomatic that the universe, brimming with billions of stars, galaxies and constellations, millions of light years distant from Planet Earth, pre-existed creation of life on earth. Nothing in Scripture suggests the contrary.

The Bible confirms angels pre-existed creation of life on earth

Job 38:7 describes the moment "when the morning stars sang together and all the angels shouted for joy" celebrating the creation event. Augustine cites a "super-celestial society," as existing prior to the creation of humans.

The Scripture makes no claim that the entire universe came into existence during the week God created earth's ecosystem. With the sun created to sustain life and to compute time, combined with the moon's tide-pulling magic, the Solar System took its place in space on day four of creation week.

© Mike Norton

The phrase "he made the stars also" can be understood to mean a reference to God's eternal power and a recognition that the stars in the universe existed already. Given the relation of the North Star to earth, the phrase could also mean that stars visible to earth were created on day four.

Earth's axis points northward, within one degree of Polaris. The North Star would be a logical component of the fourth day creation event, along with the multitude of visible stars within its sphere of influence. Polaris, with the array of stars seeming to spin around it, provides travelers with dependable directions while complementing the sun's cosmic time machine.

Like a beacon shining from what appears to be a relatively fixed position in the northern sky, this brightest star in the Ursa Minor constellation provides navigators a dependable GPS (Global Positioning System). Searches for other stars in the cosmos typically begin with the North Star.

Prior to the United States Civil War, runaway slaves looked heavenward for directions, encouraged to "follow the drinking gourd" to freedom. With this mathematically complex system in motion at the close of day four, the stage was set for the creation of animals and the human species.

Perception of time involves something vastly more significant than an abstract issue providing intellectual fodder for debate. The Big Picture significance of life is the on stage feature in the theater of the universe.

Why would the infinite, all-powerful Creator set in motion an easy-read time calibration tied to earth's orbit circling the sun, and then inspire an account of origins that discarded real *days*, leaving human imaginations to decipher the abstract?

God placed humans on a pedestal of privilege, positioned to enjoy a relationship built on regular and direct communication with the Creator. More than a superficial article of philosophical faith to accept or reject intellectually, real religion opens the heart and soul to a relationship that permeates and enriches every fiber of the being.

Creation week's day seven sets aside a one-day-in-seven celebration that promotes that relationship by helping us remember that God is the source of all life and that humans did not evolve accidentally from some "warm little pond."

The fourth commandment of the Decalogue set aside the weekly day of rest from physical labor by reminding an emancipated multitude that, "you were slaves in Egypt" and that "the Lord your God brought you out of there." [10]

The weekly rest cycle guarantees a barrier to involuntary subservience with the guarantee that all humans are created equal. Once evil intruded by corrupting creation and subverting the weekly memorial honoring the Creator, exploitation of the labor of other humans emerged as a fact of life, evolving a shackled bondage that infected and destroyed cultures.

Charles Darwin's demeaning of what he termed "savage races" provided political cover for exploitation of the barbaric slave trade. The industry's callous tentacles spanned the Atlantic, unloading manacled human cargoes in New World sanctuaries. The brutal virus carried a bitter price, above and beyond the ruthless disposal of human lives on public auction blocks.

Darwin didn't start the American Civil War. But the fall-out from his theory offered faux cover while contaminating core human values and disgracing history! The savage excesses of the slavery curse lingered to eventually rip the heart from the fledgling American republic.

Two years after Darwin published *Origin of Species*, Thaddeus Stevens, Vice-president of the Confederacy, touted slavery as "the proper status of the negro in our form of civilization…the negro is not equal to the white man; that slavery, subordination to the superior race, is his natural and moral condition." [11]

As a consequence of slavery's illicit traffic, 620,000 Americans of all colors lost their lives in the searing anguish of a Civil War blood bath. During Reconstruction, an influential cadre of white political figures, still smarting from the suppression of the rebellion, expressed resentment in pointed terms that might have made Darwin blush.

"A superior race is put under the rule of an inferior race…The white people of our State will never quietly submit to negro rule. This is a duty we owe to the proud Caucasian race, whose sovereignty on earth God has ordained." [12]

Words of hate presaged a campaign of intimidation, terror and murder.

Despite human tendency to load the weekly celebration of the creation miracle with burdensome ritual that tends to diminish its meaning, Christ challenged such onerous tradition by reminding believers, "The Sabbath was made for man, not man for the Sabbath." [13]

More than a memorial to God's power and authority to create all things capped with designing mankind in His own image, the Sabbath celebration points to the possibility of a miraculous, *"born again"* spiritual transformation through the risen Christ. It also promotes remembrance that the same Christ, the Creator, will return again to earth to make all things new and to reward all who believed his promises with life eternal.

Deep time caught the attention of theistic evolutionists offering lip service to a Creator while simultaneously dismissing the Genesis account of creation week as mere metaphor, laced with symbolic spiritual messages. Theistic purveyors of compromise dismiss literal days of creation week as "symbolic;" insisting the story of a flood "covering the entire globe" was less that worldwide. Their Christian allegiance is polluted with junk "science" rather than absolute faith in the Scripture's "thus saith the Lord."

The dismissal of Genesis creation week's literal days as *symbolic, allegory,* or *metaphor* mocks the majestic beauty of the event. Arbitrarily scrambling natural time landmarks as symbolism, merely to accommodate evolution theory, diminishes the Creator's role and defaces the big picture of life's purpose and meaning. Proposing that creation's days symbolize open stretches of indeterminate time undercuts the essence of the weekly rest day.

Theistic religionists attempt to bridge an unbridgeable gap.

© Mike Norton

Swallowing junk science bait, theistic evolutionists trash the biblical narrative of creation by injecting a disconnect between the tragedy of evil's origin and Christ's plan of redemption with its promise of life eternal.

One theistic evolutionist asserts "there is no doubt that life (and death) has been on this planet for millions of years;" while avowing evidence exists "for the presence of humans over at least 100,000 years." [14]

Attempts to mix evolution conjecture with Christian faith substitutes academic scam for Scriptural authority. Its illogical to suggest the Creator put the sun's observable time measurement in place only to play a game of "gotcha," confusing humans with a Genesis account identifying the creation week *days* as abstract *metaphors* or *symbols.*

In a search for a cosmetic commonality with Scripture, theistic evolution swallows schlock science and discards Biblical supremacy for a compromised

Christian faith, adorned with slick-sounding phraseology! Disciples of theological Pied Pipers have been hornswoggled!

It's an either/or choice. Theistic evolution mischaracterizes the all-powerful Creator of the universe and Author of all true science as a fabricator of benign myths and pious platitudes.

The Sabbath-day rest described in Exodus 20:8-11 identifies a literal, weekly 24-hour day of rest to honor the Creator and to celebrate the miracle of life. Deuteronomy 5: 12-15 restated the command to rest as a weekly reminder of Divine deliverance from dawn-to-dusk, slave labor.

The Deuteronomy reference replicated the Exodus command to rest one literal day each week, correlating with the fourth commandment of Exodus and the Genesis account of a seven-day creation week. If only metaphor, God's guarantee of a literal day of physical rest each week would offer little comfort to exhausted former slaves who survived the searing heat of Egyptian "iron-melting furnaces." [15]

It makes no sense to suggest the 24-hour day earmarked for physical rest and celebration is a metaphor for an abstract stretch of time longer than a human life. Promising "symbolic" rest from physical labor spouts hollow rhetoric, and a distortion of obvious, Biblical meaning.

Demoting the Divinely ordered weekly rest day to "symbolic" status, corrupts its significance and subverts God's authority. Why would Genesis creation *days* be metaphor while Deuteronomy and Exodus memorialize creation with weekly literal-time rest *days*?

Christ reassured believers that "The Sabbath was made for man." [13]

Finite minds struggle to comprehend and calibrate infinite time and space. Concepts of time without beginning or end, and cosmic space without boundaries, staggers mortal intellects.

Human perception of time in the context of days, weeks, and years correlate with the Scriptural account of beginnings.

XVI

Time Zero

Young Life, Old Earth

"Ultimately, the Darwinian theory of evolution is no more nor less than the great cosmogenic myth of the twentieth century." [1]

Michael Denton

© Warren L. Johns

With evidence of evolution a virtual no-show during civilization's known history, the theory has no place to go other than to deep time's mystical scenario.

When Washington State's Mt. St. Helens blew its top, torrents of hyper-heated ash engulfed and destroyed wholesale quantities of plant and animal life, including *Homo sapiens.*

Where a proud mountain peak once dominated, a concave chasm confirmed the reality of nature's catastrophic reach, exploding on a date certain---8:32 a.m., May 18, 1980. Human casualties from the explosion included Harry Truman (an old time mountain man, not to be confused with a former U.S. President), who ignored warnings. His vaporized remains have never been seen since.

While the real-time record of the 1980 event stands undisputed, initial radiometric data of the lava dome conflicts, out-of-kilter.

"Rocks formed in and subsequent to the 1980 eruption ...should date 'too young to measure'...According to radioisotope dating, certain minerals in the lava dome are up to 2.4 million years old.

"All of the minerals combined yield the date of 350,000 years by the potassium-argon technique," despite the fact, "these minerals and the rocks that contain them cooled within lava between the years 1980 and 1986." [2]

A 350,000 to 2.4 million-year time stretch shouts discordance. The discrepancy highlights a fossil-dating dilemma. It's certifiable fact mountain man Harry Truman, lost his life in May, 1980. Thus, just because the radiometric age allocated the turf surrounding his remains is dated 350,000 or 2.4 million years before the present, doesn't mean Truman lived and died 350,000 or 2.4 million years ago. Harry Truman lost his life in 1980, regardless of the projected radiometric dates assigned to the fall-out residue from the volcanic event of that date.

This reality wreaks havoc with evolution's dubious technique of awarding fossils the radiometric date of the matter composing the burial site.

© Wikepedia

Mt. St. Helens following its 1980 eruption.

Fossils of plant and animal life forms that lived and died several thousand years before the present could be buried in matter showing radiometric dates millions of deep time years BP. It's less than logical to equate a plant or animal date of death with the radiometric age of the matter composing its burial site. Radiometric dating of inorganic matter to determine the age of the fossil remains of an organic life form ignores the fact fossils don't automatically assume the age of the surrounding burial site turf.

No question about it, both earth's inorganic matter and its plethora of organic life forms had a *Time Zero* beginning. Genesis describes a lifeless, inert water-covered matter, floating in dark space at the beginning of creation week. "Now the earth was formless and empty, darkness was over the surface of the deep, and the Spirit of God was hovering over the waters." [3]

If we take the Genesis narrative to mean what it says, then God created all plant and animal life, with a supporting ecosystem, during the first six days of a literal seven-day week.

No reference is made to the creation of water or that inorganic raw material was in place at the beginning of the week God created life with its supporting ecosystem. Nor does the Bible define the length of time that water-covered matter floated in cosmic darkness prior to creation week.

One observer, not a Bible scholar apologist, cites "evidence showing that, 3.5 billion years ago, Earth was mostly covered by water." [4]

"Formless and empty" matter provided the foundation upon which God created life---original, unborrowed and underived. An already existing universe celebrated the event "When the morning stars sang together, and all the sons of God shouted for joy." [5]

Some creationists dispute the pre-existing universe reality, conjecturing the recent creation week encompassed the entire cosmos. No persuasive evidence, either from science or Scripture, suggests the recent week life was created, several thousand years ago, marks the birth of the universe.

This claim doesn't jibe with the reality that located "in one of the Milky Way's outer spiral arms…the Sun lies between 25,000 and 28,000 light years from the Galactic Centre." [6] Some estimates suggest our boundless universe extends 100-billion light years distant.

Such calculations, if valid, clash with assertions of creationists who believe the total universe was created the week life was created on earth. Its not consistent to allege a creation event that may have occurred as recently as 6,000 years before the present when persuasive evidence argues for a universe so vast that, even at the speed of light, there is not nearly enough time for many stars in our own Milky Way to be yet visible on earth.

The implication is clear: several thousand years before the present, there was a *Time Zero* moment God's command introduced life, together with a global ecosystem, that miraculously transformed the face of formless matter into the Blue Planet, throbbing with young plant and animal life.

Physicist Robert H. Brown interprets the Bible account to mean what it says: a lifeless mass, inundated with water, cloaked in a blanket of darkness. He describes the possibility of "radioactive decay over hundreds of millions of years prior to Creation Week" as feasible, taking Genesis 1:2 to mean what it says.

"New Testament writers make it absolutely clear that all components of the physical universe were created by God, but do not specify a time frame. The only necessary basic time specifications are provided in the first 11 chapters of Genesis. And those specifications apply only to living organisms and the environment that supports them.

"Nothing is said in the Bible about time in connection with the creation of water or the creation of the inorganic material that was raised above surface water on the third day of Creation Week---these components are simply stated as being 'there' at the beginning of Creation Week." [7]

The Bible doesn't express or imply symbolic "days" or gradualism.

Twentieth century biologist Frank Lewis Marsh suggested all earth's life forms necessarily would have been created in a fully mature format, imbued with an appearance of age, thriving as components of a complete, fully functioning, irreducibly complex ecosystem. "Our earth was created, along with the living forms with which it was furnished, with an appearance of age." [8]

Mature trees stretched toward the sky, fish swam, birds flew, flowers bloomed, vegetation flourished and animals roamed.

Adam did not arrive as a helpless infant, but walked the planet as a fully-grown human, caring for himself and capable of reproduction. He required nourishment to sustain life, not just seeds to sow. Food crops grew in place, ripe and ready to eat. Without plants and trees bursting with fruits and vegetables, Adam would have starved.

The volcanic island of Surtsey, just south of Iceland, first surfaced the Atlantic Ocean, November 14, 1963. A casual observer might mistake the landmark to be several thousand years old. Less than four years after its appearance, a paleontologist flew to the site to inspect the pristine landscape, later reporting its appearance of age.[9] He saw a brand-new chunk of raw geography projecting a weatherworn appearance. A relentless sea had ground out black sand beaches. Multi-layered cliffs, composed of a series of lava flows, guarded a four-mile coastline, carved methodically by Atlantic tides. Foot-long stalactites hung from cave ceilings. Basalt blocks appeared as rounded boulders, chiseled by the elements. Despite its tender 1963 age, sea gulls, insects, and at least three species of plants already called Surtsey home.

God is not limited to a single creative event in the context of infinite time. Christ, the Creator,[10] possesses the power to create earth's life forms in real-time days. Anything less and He would not be the all-knowing, all-powerful, all-loving God, worthy of human worship.

When humans attempt to recreate God in their own image, they distort, disparage and diminish the majestic power and authority of the one who created all things.

Aurelius Augustinus (Augustine, 354-430 A.D.), Bishop of Hippo and Christian intellectual, is believed to have played an influential role in the African Synod of Hippo (393 A.D), which ratified the joining of the Old and New Testament components for the biblical Canon. Prolific writer and visionary Christian leader, Augustine quoted liberally from both Old and New Testaments. Committed to the integrity of the creation miracle, Augustine attributed the Genesis narrative to the pen of Moses writing under God's direct inspiration.

The words of Moses, a commanding presence with giant intellect and charismatic leadership skills, warrant serious attention and sober respect. In today's academic era, which insists on the pre-eminence of primary sources, the unmatched credentials of Moses tower tall. He grew up in the privileged environment of an Egyptian royal family ruling an empire that straddled ancient world trade routes. On at least two occasions, at the burning bush and on Mt. Sinai, Moses communicated directly with God, learning the Genesis narrative from the Creator.

Level granite surfaces in Egyptian pyramids were trimmed to a tolerance of $2/1000^{th}$ of an inch; deep rectangles were carved with square corners; and circular holes were drilled through the face of solid rock leaving geometrically precise, cylindrical openings.

© Sculpics

Egypt's Giza Pyramids

The tool design and the millenniums-old technology inspires wonder. "It may be…that the science we see at the dawn of recorded history was not science at its dawn…[but] remnants of the science of some great… civilization." [11]

With the background of an impressive Egyptian culture that survives antiquity, higher education would have been unsurpassed with its records of ancient and contemporary civilizations carved in stone and painted on rolls of papyrus. More than having been indoctrinated with state-of-the-art secular wisdom and historical fact, Moses learned the culture of the Hebrews at the feet of his mother, Jochebed.

The pivotal life and times of Noah would likely have warranted front-and-center billing in the education of Moses. As a transitional figure of living faith, Noah walked and talked with humans capable of sharing eyewitness accounts of events that occurred centuries earlier---before, during, and after the global deluge. Methuselah, a descendant seven generations down from Adam, enjoyed a life span that would have overlapped with the lives of both Adam and Noah.

Moses was the beneficiary of a cultural narrative built on both written records and eyewitness accounts. In contrast to the vagaries of tradition, Moses' words reflect the "best evidence rule."

Already a powerhouse source of ancient history, Moses's credentials and writings were authenticated under the inspiration of the Creator Himself. From the "burning bush" communication to Mount Sinai's summit, Moses wrote with authority, including the Genesis summary of creation week.

Writing in 1654 A.D., James Ussher, authentic scholar and Christian clergyman, proposed creation week to have occurred in 4004 B.C. based on his interpretation of Genesis genealogy. Recent creationist scholars tend to believe several "thousand" years might be added to Ussher's estimated 6,000 years before the present but not "millions."

Since the time lapse from creation week to man's fall and eviction from the Garden of Eden is not defined in Genesis, a speculative question arises: Was Adam's age calculated from the moment of his creation or after being barred from access to the tree-of-life with onset of his physical aging?

Humans don't know the answer nor is it necessary to know. Whether or not the Bishop Ussher chronology offers the last word in dating the creation of life on earth, above and beyond his academic skills and the authenticity of his primary source material, there is evidence in both nature and archaeology suggesting life on earth originated only thousands, not millions of years in the past.

For a variety of reasons, Ussher's genealogical scenario is entitled to thoughtful evaluation, not to be dismissed out-of-hand.

For one thing, the scholarship and integrity of Moses deserves respect. For another, the extended lifetimes of humans from Adam through the pre-global flood era lends plausibility to eyewitness accounts descriptive of events both before and after the worldwide deluge.

Recognizing the brilliance and longevity of ancient humans, we can propose that it is likely written records in some format were put in place in order to preserve the history of noteworthy events both before and after the flood. Apart from slightly different time variations offered by the Septuagint version of Genesis, one possible deficiency of Ussher's chronology, is that the Hebrew word "ab," translated "father," can also mean "grandfather" or "ancestor." But that can be argued either way.

For example, Genesis 10 describes Shem as "...the father of all the children of Eber," while Genesis 11 confirms a two generation gap between Shem and Eber, indicating the word translated "father" can mean "ancestor."

While Ussher never claimed infallibility, his carefully crafted analysis of the Genesis genealogical record combines a perspective of antiquity with an insightful look at the big picture meaning of human life. Skeptics, questioning Ussher's thesis, point out that genealogy is a less than an adequate substitute for chronology. Still, critics have nothing remotely comparable to offer.

Archaeology and natural science provide glimpses of antiquity. Measured in thousands and not millions of years, recorded history delivers no evidence demonstrating evolution-in-action.

Whether or not the Ussher chronology offers the last word in the dating of creation of life on earth, there is evidence in both nature and archaeology that deserves consideration in support of his analysis.

Evidence of plant life reaching back 4,000 years before the present tracks with Ussher's dating of the global flood to 4300+ years ago. Plant and animal life, virtual mirror images of today's fauna and flora, flourished without a hint of evolution in action.

"Methuselah," a scrawny bristlecone pine clinging to the rugged slopes of the Sierra Mountains, survives today, rooted in more than 4,000 years of earth's immediate past. A Giant Sequoia carries rings confirming its life began 3500 years before the present. Egyptian mummies represent four *Homo sapiens* blood types, identical to what flows through the veins of twenty-first century humans.

"The fact that even the Great Pyramid at Giza, which according to modern historians is only supposed to date from approximately 2,600 BC, can contain such a remarkable wealth of a mathematical, astronomical, and the geophysical information (it not only contains the value of pi, and is aligned with Sirius, but it's dimensions accurately describe the major measurements of the Earth) indicates that there was an immense scholastic tradition in Egypt by 2600 BC...

"Even if we were to assume that ancient Egypt, like the Druidic priesthood, relied to a large extent on oral tradition, the fact of architectural construction, mathematical ingenuity, and astronomical prediction and calculation, indicates a library tradition to both compute, research, and convey lore, knowledge and wisdom from generation to generation." [12]

These projected dates correlate comfortably with the date of the global flood as calculated by the Bishop's Biblical-based chronology.

The Bible reports that following the flood, descendants of Noah, speaking the same language, congregated on the Plain of Shinar, intending to build a tower tall enough to spare them from the devastation of future floods. To halt construction and to disperse the population, God confused "their language" so they could not "understand each other." [13]

Subsequently, diverse written languages were developed geographically. The Hebrew alphabet originated at least 3500 years BP. [14] Between the years 2100-1500 BC, at least five scripts were known to be in use, including Egyptian hieroglyphs, Acadian Cuneiform, the hieroglyphiform syllabary of Phoenicia, the linear alphabet of Sinai, and the cuneiform alphabet of Ugarit. [14]

Standout civilizations with visible archaeological roots reach back to the Sumerian, 3500 BCE; the emergence of Egyptian power by 3000 BCE; and Indus Valley stirrings as early as 2600 BCE.

The Chinese calendar boasts antiquity. The reign of Yu, the first emperor of the first Chinese dynasty (Xia), arose BC 2205.[15] The Shang Dynasty's written history surfaces 1766 BCE.

These landmark time frames represent composite touchstones for the early beginnings of modern history---all since 4349 BC. Evolution is a no show throughout this entire modern history era.

If eight humans survived a worldwide deluge only thousands of years before the present, skills in working with iron, bronze and copper would have to be reintroduced. Instead of bumbling cave-man types grunting out their emotions and surviving on brawn, global landscapes are loaded with visible marks of sophisticated minds left for discovery.

Although ^{14}Carbon dating doesn't assure slam dunk precision, some believe civilization's beginnings appeared at least 5500 years BP.[16] Because of the ice age's receding edges in Europe, archaeological tracking of recent history reveals traces of human habitation in Jericho, adjacent to the Mediterranean Sea, 10,000 years BP. In today's Turkey, remnants of refined ancient architecture still stand. [17]

From evidence he evaluated, Charles H. Hapgood, author of *Maps of the Ancient Sea Kings*, believed that far from the imagined successive stages of development from traditionally touted primitive eras of Stone, to Bronze, and to Iron--ancient cultures mastered mathematics, astronomy and science. Granite monuments from known, past civilizations confirm that human ancestors were anything but knuckle-dragging oafs who communicated with guttural grunts. Astounded by the evidence of advanced technology in old times, some of those buying into the questionable IQ level of original humans wonder if "aliens" provided the know-how.

After completing analysis of a portfolio of ancient maps of the world indicating use of spherical trigonometry, Hapgood, concluded, "in remote times, before the rise of any of the known cultures," there well could have been "a true civilization, of a comparatively advanced sort…even more advanced that the civilization of Egypt, Babylonia, Greece and Rome…

"The mapping of a continent like Antarctica implies much organization" and "many exploring expeditions…It is unlikely that navigation and mapmaking were the only sciences by this people." [11]

A mystery lingers as to the source of relatively accurate, ancient maps that appear to document portions of Antarctica's coast and its continental landmass before it was covered by its present ice cap.

"These older maps were based upon a sophisticated understanding of the spherical trigonometry of map projections, and---what seems even more incredible---upon a detailed and accurate knowledge of the latitudes and longitudes of coastal features throughout a large part of the world." [11]

Fossil evidence confirms that sometime in the long ago, today's ice-loaded Antarctica provided a home for a variety of plants and animals, typical of a more temperate climate.

"Abundant finds of fossil leaves and wood point to the existence of extensive forestation in earlier geological periods, even to within a few degrees of latitude of the South Pole itself.

"Dinosaurs, and later, marsupial mammals once roamed across its surface." [18]

Does this paleontological reality correlate with a relatively recent time when the *Maps of the Ancient Sea Kings* might have been charted? Is it possible that maps charting the coast of an earlier Antarctica were charted before the deluge and preserved by Noah in written archives for post-flood times? Could that data have been a reference source for the development of maritime skills that guided venturesome sailors around the globe, hundreds of years into the post-flood future?

While definitive answers are conjectural, it seems rational that human generations prior to and after the global flood understood maritime ship design and the mathematics of navigation.

XVII

Life's Slippery Slope

Devolution

"We know that the whole creation has been groaning,

as in the pains of childbirth, right up to the present time." [1]

Paul, the Apostle

© Caitlin Mira

The Second Law of Thermodynamics demonstrates its effect on this tipsy remnant of a century-old California ghost town.

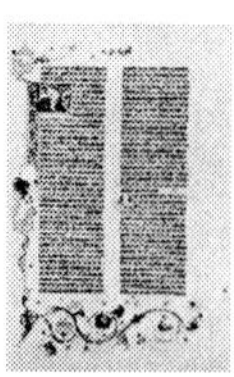

Less than fifty years after Darwin unveiled his molecule-to-man scenario, divers off the coast of the Greek island Antikythera discovered encrusted remains of a mechanical device that is still raising eyebrows in the 21st century.

Designated the "Antikythera Mechanism," the 2,000+ year-old machine with its 30 bronze gears and more than 80 fragments continues to intrigue researchers unraveling its mysteries a century after its discovery. Even before surrendering its secrets, the delicate design suggests that, far from evolution's mischaracterization of ancient humans as knuckle-dragging oafs who communicated with grunts and groans, our ancestors walked, talked, and created as fully developed *Homo sapiens.*

Wikepedia

The remains of the complex, Antikythera Mechanism

Contemporary scholars believe the Antikythera Mechanism, built circa 87 B.C, carries information showing the earth's relationship to the sun, our moon and the five planets recognized by the Greeks. X-ray technology has identified inner-workings as complex as a 19th century Swiss clock that also track the Olympic year cycle.

The technological and engineering skills, carved in stone by human ancestors, continue to amaze us. These weather-beaten monuments, thousands of years old, testify that knuckle-draggers have not been on the scene for at least the last 4,000 years. Ancestor humans lived longer and were at least as intelligent as their modern counterparts.

"Egyptian civilization was not a 'development' but a 'legacy.'" [2]

Sophisticated tools used in the cutting, polishing, hauling and placing precisely shaped multi-ton blocks of granite and marble defies explanation. Yet there they stand, architectural masterpieces for the entire world to admire. Egyptian pyramids and England's Stonehenge were not conceived and put in place by mutant amateurs.

What tools were used to cut square corners and long, perfectly straight rectangular grooves in the face of solid granite without trace of a chisel mark? How was it possible to polish giant slabs of marble as smooth as glass, free of flaws? And what about those precise, circular holes drilled into the hardest rock blocks?

Answers might have been found in one or more of those ancient libraries that no longer exist.

Asia Minor's Pergamum Library, created after the death of Alexander the Great, was the geographic center for the development of parchment. This learning center reportedly contained 200,000 items. Its needless destruction deprived future generations of access to a prized compendium of Phoenician culture, history and science.

North Africa's Carthage Library reputedly housed 500,000 items prior to its destruction by the Romans in 146 BC. And the famed Egyptian library in Alexandria, founded by Ptolemy I sometime between 323 and 283 BC, was believed to have included as many as 700,000 items (some estimates range to a million) until destroyed in 48 B.C. by Caesar's reckless legions.

Obliteration of these treasuries of ancient world history and scientific wisdom eliminated ready access to accounts of ancient times and set the stage for civilization's downward spiral toward Medieval era darkness.

History exhibits retrogression and devolution rather than "progress toward perfection."

Peer review offered no deterrent to Charles Darwin's musings. Devoid of significant academic credentials, he compiled ponderous writings based on personal opinion influenced by Grandfather Erasmus Darwin. At the age of 50, he rushed *Origin* into print, which was cobbled together from personal observations during his five-year excursion aboard the HMS Beagle and his pigeon experiments conducted at his Downe family estate. Lacking knowledge that a living cell consisted of anything more than a blob of protoplasm, much less DNA's "language of life," he promoted his ideas fully aware the preponderance of fossil evidence didn't support his molecule-to-man imaginings. Other than pitifully few, debatable finds, those non-existing "missing links" remain invisible.

The shibboleth of some "little pond," serving coincidentally as an incubator for spontaneously generating first life, smacks of flat earth mentality. Recognizing obvious "flaws" and "holes" plaguing his ideas, Darwin reached out to American Asa Gray, bemoaning the possibility he may"...have devoted my life to a phantasy [sic]." [3]

Debunked and devalued, *The Origin of Species* still survives as a calcified tribute to the powers of social status, economic privilege and blind tradition in the preservation and propagation of thin-soup thinking.

"Evolutionism, or the 'theory of evolution'...is the erroneous idea that evolution is the source of progress and the cause of the cosmos structure---life and any other systems in the universe." [4]

The fossil record abounds with evidence of once thriving, giant species. What has emerged from 150 years of discovery is a cross-section of gigantic fossil ancestors that dwarf replica descendants. No genealogical trail of half-evolved predecessors precedes the sudden appearance of these giant ancestors. Genetically complete prototypes appear fully formed and in place from time zero.

One thing is certain: a long time ago, giant species roamed earth.

The roster of jumbo-sized ancestors features a crocodile with a seven-foot head and a fossilized, cow-sized pig housed in a Denver museum. Washington, D.C.'s Smithsonian displays a sloth fossil as large as a pickup truck. Rhinoceros-sized marsupials and giant kangaroos once inhaled Australian air. Foot-long trilobites occupied what is now Morocco. Oversized lemurs hung out in Madagascar. The Caribbean's Anguilla Island at one time provided home base to a three hundred pound rat. Colorful dragonflies with inch-thick bodies propelled by two-and-one-half foot gossamer wings hovered over pools of cool water in some ancient past. Armor-plated armadillo missed Darwin's message that descendants must not only modify but also improve. Instead, today's armadillos appear to be puny versions of their nine-foot-long ancestors.

Tons of dinosaur bones, uncovered late in the nineteenth-century in the rugged high prairies of Colorado and Wyoming, testify to giant-sized life forms that no longer exist. Massive dinosaurs, some believed to have extended as long as 150 feet tip-to-tail, once roamed the earth. Twenty-foot-long great white sharks patrolling Pacific shores strike terror, but the fossil remains of an ancestor shark found on dry land near Oildale, California, weighed eight times as much, measured forty feet in length, and sported a twelve-foot head.

Foothills high in the Himalayas "contain fossil beds rich with extinct terrestrial animals, a tortoise twenty feet long, a species of elephant with tusks fourteen feet long, and three feet in circumference." [5] And speaking of fearsome, how about "…The monstrous Carcharodon… possessing distinctive triangular teeth up to 8" long, [that] may have had a 6-7ft wide jaw gape, and a length of 80 ft." [6]

Australia provided homes for other now-extinct giants. Imagine a tortoise the size of a Volkswagen Beetle, a Kangaroo ten-feet tall, or a wombat-like creature the size of a hippo.[7] The enormous Australian bird, *Genyornis newtoni*, makes the twenty-first century emu appear scrawny.

Elephant-sized ground sloths roamed South America.

The eye-catching list of giant-sized fossils includes a fourteen-inch tarantula... a millipede-like creature six feet long and a foot wide... giant beavers... ammonites several feet in diameter... *eurypterids* (enormous crabs), ranked among the largest invertebrates ever... pterosaurs with wingspans reaching fifty feet... canary-size mayflies.

Fossil piranhas have been found four times the size of their ravenous descendants. Imagine the pesky cockroach, equipped with an intimidating foot-long body to better perform its stealthy mischief.

Think about fossil clams spanning 12 feet across, weighing in at 650 pounds, discovered in the Andes 13,000 feet above sea level in the Huancavelica province of Peru. [8]

A jellyfish 28 inches in diameter that "...must have been buried extremely rapidly..." was discovered in a Wisconsin "fossilized beach." [9]

Mosses 2-3 feet high (compared to today's 1 to 3 inches) brightened the landscape; horsetail reeds grew up to 50 feet tall (ten times today's 5 to 6 footers); and a hornless rhinoceros towered 18 feet high while stretching 30 feet in length.[10]

Next time you explore California's coast, you might gaze up the trunk of the Grizzly Giant, one of a grove of still surviving descendant Sequoia trees, still stretching nearly three hundred feet into the wild blue yonder.

Don't overlook the over five-foot long coiled shellfish (compared to 8 inches today); the eight foot wide bison skull on display at the Mt. Blanco Fossil Museum, Crosbyton, Texas; the ten-foot tall ammonite shell on display in a German museum; and turtles nearly four meters long.[11]

The Island of Minorca has housed the fossil bones of "a huge bat, a large mouse, and a giant tortoise" and a rabbit reported to be "six times as big as a rabbit today." [12] Enough to make the mythical Bugs Bunny hang his head.

Finally, check out the eight-foot shell of the giant nautiloid (resembles a modern squid), [13] and the fossil footprints in Canadian sandstone believed to be those of a twenty-inch long centipede, five times the length of its mini-sized descendants.[14]

This list rambles on, but you get the picture.

Far from isolated freaks of nature, giant life forms highlighted a biodiversity that no longer exists as a common sight in century twenty-one.

Life on earth decays and deteriorates.
Wrinkles etched in aging human faces, rusting iron,
and crumbling concrete characterize a parade of devolution.

Darwin imagined that some simple, reproducing life form, created itself accidentally from non-living matter, in some "warm little pond," millions of years ago. Then, incrementally, supposedly, it randomly transformed itself into every kind of plant and animal life that inhabits today's earth. Ever the optimist, Darwin relied on the thought processes of his own mind to piece together his "progress toward perfection."

Nothing could be further from the truth. The concept remains a tortured paradox.

Evolution theory came from a human brain that, if true, would have fashioned itself accidentally over eons of time through a series of trial-and-error transits. Reality points instead, to all life on earth having been designed and created perfect, pristine and pure, at the top of its game, by the command of a *"Superior Rationality,"* the Lord God Almighty.

From a simple cell creating itself accidentally in some "warm little pond," evolution mischaracterizes human ancestors as stooped, knuckle-draggers, gradually evolving upright strides.

Ancient *Homo sapiens* were nothing of the kind. The Scripture narrative attests to a bucolic beginning when giants thrived and humans, created in God's image, lived long lives in an environment of perfect biodiversity.

When the first *Homo sapiens* couple fell victim to deceit and chose disobedience, perfection slipped into devolution's downhill pattern of decay and eventual death.

The oldest surviving written record describes oversized *Homo sapiens.* 'The Nephilim were on the earth in those days...They were heroes of old, men of renown." [15]

Years after the global flood, ten of the twelve Israel spies sent to evaluate defenses in the land of Canaan, returned with a discouraging report, again citing the Nephilim.

"They are stronger than we are...All the people we saw there were of great size... We saw the Nephilim there...We seemed like grasshoppers in our own eyes, and we looked the same to them." [16]

© Steven Castro

Shiny new cars inevitably succumb to rust and corrosion when abandoned to the whims of mother nature.

Absent intelligent intervention, devolution, not evolution, rules. The human species is not immune from the insidious impact of devolution.

Even Disneyworld's magic would be overgrown with weeds, worn out mechanisms and pealing paint without 24/7 maintenance.

Original life perceptions represent diametrically opposing views. The Second Law's impact contributes to ecosystem decline, inflicting its relentless toll on plants, animals and all material things.

The Second Law of Thermodynamics erodes the Earth's ecosystem.

The label "evolution" has been high jacked when used to connote upside progress. The term "devolution" more accurately reflects reality. Upside-down rhetoric saturates a culture intent on ignoring the Creator.

Purveyors of Neo-Darwinism commit to a materialistic faith that credits an accidental "act of nature" as the source of life. But when those same natural forces mar the landscape, knee-jerk semantics mislabel the disaster an "act of God."

Earth, with its teeming biodiversity, should be attributed to the creative act of an all-powerful God, deserving worship, while the destructive forces of hurricanes, earthquakes, floods, and tornadoes should be recognized as acts of rampaging nature.

According to the First Law of Thermodynamics, energy can neither be created nor destroyed. The Second Law of Thermodynamics is qualitative, confirming the tendency of energy to flow away, to disperse from concentration in a localized place. Entropy is the quantitative measure of the dispersion or spreading out of the qualitative. It represents a measure of change of energy distribution after some spontaneous event.[17]

Unlike gold that won't stain or tarnish, iron, steel and concrete suffer the dragon's breath of corrosion. A century after it's founding, Rhyolite, Nevada's bits of broken buildings, testify to the Second Law's triumph.

Forces of wind, water, snow and ice are wearing down the world's loftiest mountain chains. Grinding glaciers gouge out scenic valleys, carved midst the rocky terrain of jagged peaks, once sculpted by hydraulic action.

Mathematician Granville Sewell describes the dilemma confronting evolution when scrutinized under the laws of physics, mathematical probability and the Second Law.

© Warren L. Johns (Duluth, MN, 1951)

Albert Woolson, last survivor of Abraham Lincoln's Grand Army of the Republic, lived to be 109 years old-- a fraction of the longevity of our Old Testament ancestors.

"The underlying principle behind the second law of thermodynamics is that natural forces do not do extremely improbable things...natural forces do not do macroscopically describable things which are extremely improbable from the microscopic point of view...

"The second law predicts that, in a closed system where only natural forces are at work, every type of order is unstable and will eventually decrease, as everything tends toward more probable (more random) states---not only will carbon and temperature distributions become more random, but the performance of all electronic devices will deteriorate, not improve.

"Natural forces, such as corrosion, erosion, fire and explosions, do not create order, they destroy it. The second law is all about probability. The reason natural forces may turn a spaceship into a pile of rubble but not vice-versa is probability: of all the possible arrangements atoms could take, only a very small percentage could fly to the moon and back." [18]

The organic world can't escape the Second Law's relentless effect.

Any 80-year-old understands time's ravages. Once robust and graceful twenty-year-old bodies tend to falter and bend; blotches mar once-radiant skin that dries, wrinkles, and folds; eyesight and hearing require artificial boosts; hair thins into wisps of white and gray while mental marbles roll in slower motion as memories falter, then fade.

Cosmetic surgery may camouflage deterioration, but an undeterred Second Law of Thermodynamics delivers the last word. Sooner or later the all-powerful Second Law touches all lives. Daily exercise, heavy doses of vitamins, and impeccable genes won't prevent the most beautiful young woman or the most muscular young man from deteriorating, cell-by-cell.

Life slips into the slow lane as systems drift downhill. Humans confronting warning signs of aging understand the approach of their inevitable date with the grim reaper, no longer deferred by a once formidable immune system. In time, the heart stops beating and black curtains shroud a cold-dust destiny. Humans can split atoms but can't create them, much less design and build a living cell.

Nature scoffs at Darwin's "progress towards perfection" cliché.

Neither man nor matter escapes the ravages of decline. Perfection reigned at the close of creation week. But when humans fell for the arch deceiver's assurance that by tasting the forbidden fruit they "would not surely die" but "your eyes will be opened, and you will be like God," [19] devolution took control.

Could it be that some strange, mongrel style life forms observed in nature resulted from devolution, aided and abetted by debilitating mutations? Eyebrow raising speculation offers no answer. But it is beyond speculation that the Second Law of Thermodynamics, the "supreme" law of nature, erodes and ultimately shatters perfection.

Wise ancient writers addressed nature's inexorable downside trend. Citing the heavens and the foundations of the earth, the Psalmist predicted: "they will all wear out like a garment." [20]

XVIII

Mother Earth's Facelift

Global Gully-Washer

"Areas of unknown extent are buried under strata which rest on them uncomformably, and could not therefore constitute the original capping, under which the whole of these rocks must once have been deeply buried."[1]
Alfred Russell Wallace

© FloridaStock

Cathedral Rocks exhibit colorful clues to nature's power.

Genesis describes a global-wide, hydraulic catastrophe in which "waters flooded the earth for a hundred and fifty days" covering "all the high mountains under the entire heavens." [2]

At the beginning of creation week, Genesis describes a water-inundated earth "without form…" with darkness "upon the face of the deep…." [3]

Donald R. Lowe, Stanford professor of geological and environmental sciences, with colleague Louisiana State geologist Gary R. Byerly, speculate that 3.47 million years ago, earth may have been mostly covered with water when struck by a meteor. The scientists reasoned if the ocean contained the same volume of water as today, it would have been two miles deep. [4]

If true that water covered the earth prior to the creation of life, it's not unreasonable to consider the possibility of a relatively recent worldwide hydraulic deluge covering "all the high mountains" after life's appearance. One twentieth century observer believes "the world got soaked seventy million years ago, as sea levels rose five hundred feet." [5] The "five hundred feet" assessment sounds suspiciously global, at least partially reminiscent of the Genesis account of the deluge. A defense witness in the 1925 Scopes trial said "practically all of the earth has at some time or other been covered by water." [6]

So, if earth's land mass can be flooded in piecemeal fashion, why not a global gully-washer sweeping the landscape in one fell swoop?

Even Chile's Atacama Desert, reputedly the driest strip of land on the planet, is said to have been a wetlands at one time, covered with residual water from the ice age, 11,000 years before the present.

Diverse cultures perpetuate a variety of traditions describing a global hydraulic event that swept earth's geography in ancient times. While none of these hand-me-down stories resemble the details spelled out in

Genesis 7, the collective references add credence to a cataclysmic hydraulic event that permanently scarred the face of the Earth.

The Gilgamesh Epic gives details that quite closely resemble the Genesis account of the global flood. Some skeptics argue that Moses copied portions of the Gilgamesh legend.

Alternate contrary arguments might be made that either the secular Epic plagiarized from Noah's eyewitness record of the flood, or else it's 12 tablets simply represent after-the-fact, multi-edited compilations of hand-me-down traditions of a very real, world-shattering event.

The Akkadian version discovered in Ashurbanipal's Ninevah library, is believed to have been dated sometime between 1,300 and 1,000 B.C. Moses, a charismatic presence blessed with a commanding intellect, lived and wrote earlier, preserving the primary record of the unprecedented hydraulic action in his Genesis narrative.

More than three millenniums later, Charles Darwin rejected the idea of a worldwide flood; it didn't track with his dream. The English naturalist asserted without equivocation that "we may feel certain…no cataclysm has desolated the whole world." [7]

We are not privy to the source of Darwin's "inside" information disputing the worldwide flood, but mountains of data suggest that he didn't know what he was talking about.

What Darwin neglected to explain, with his out-of-hand rejection of the Global Flood, was just how the un-renewable coal and petroleum deposits, along with scattered cemeteries of disarticulated fossils, were compacted, buried suddenly and scattered throughout "the whole world" by water born sediments. Nor did he account for the mix of land animal fossils, such as the dinosaur M*uttaburrasaurus* in Australia and *T-Rex* in Canada, "found buried with marine creatures such as shellfish, turtles and fish…" [8]

While Darwin's five-year HMS Beagle adventure exposed him to exotic locations, he never came close to circumnavigating and investigating the entire world. Despite commendable self-taught investigative research, the naturalist lacked academic credentials as a geologist, hydrologist or paleontologist. Nor did his five-year excursion aboard the Beagle provide sufficient data for his arbitrary judgment call dismissing a worldwide flood.

Turning his back on the biblical account of life's creation miracle, he must have understood that if he bought into the deluge of Noah's day, his thesis would be hung-out-to-dry, demolished and swept away as so much fossilized flotsam in a rising tide of contrary evidence.

Seventy percent of earth's surface is water covered. The remaining thirty percent is etched with watermarks. Localized floods and natural disasters have also repeatedly marred and scarred that thirty percent. Alfred Russell Wallace, evolutionist contemporary of Darwin and hardly a fan of the global flood of Noah's day, cited "denudation" and "destruction "of the earth's geology as explanation for the lack of transitional forms. However inadvertent the reference, his description of earth's geology can be read to imply markings of a past worldwide deluge.

"...Denudation is always going on, and the rocks that we now find at the earth's surface are only a small fragment of those which were originally laid down...the frequent uncomformability of strata with those which overlie them, tell us plainly of repeated elevations and depressions of the surface, and denudation on an enormous scale.

"Almost every mountain range, with its peaks, ridges, and valleys, is but the remnant of some vast plateau eaten away by sub-aerial agencies; every range of sea-cliffs tell us of long slopes of land destroyed by the waves; while almost all the older rocks which now form the surface of the earth have been once covered with newer deposits which have long since disappeared."

Wallace opined that "...areas of unknown extent are buried under strata which rest on them uncomformably, and could not therefore, constitute the original capping under which the whole of these rocks must once have been deeply buried; because granite can only be formed, and metamorphism can only go on, deep down in the crust of the earth...What an overwhelming idea does this give us of the destruction of whole piles of rock, miles in thickness and covering areas comparable of those of continents..." [1]

In this sense, evolutionists and creationists agree: Mother Earth's face has undergone drastic revisions over time.

Evidence of waterpower's destructive revenge can be seen just east of Seattle, where a colossal water surge sculpted a 16,000 square mile gouge in earth's surface in the blink of an eye.

Investigating geologist J. Harlan Bretz raised eyebrows when he suggested to professional peers in a 1927 lecture to the Geological Society of Washington, D.C. that the State of Washington's Scablands' scar didn't result from an eroding river or a grinding glacier. His radical hypothesis that such a massive swath gouged the earth suddenly, thanks to a mega-surge of water, scandalized professional peers. The listening audience of geologists dismissed the suggestion out-of-hand because it carried biblical implications. The Bretz idea flew in the face of conventional thought, which was committed to the idea that markings of this magnitude required millions of years of erosion by a river or a glacier.

It took a lifetime of analysis for the revolutionary Bretz interpretation to gain favor. But more than a half-century after the fact, recognition finally arrived in a ceremony tied with symbolic ribbons. In 1980, geologists honored the aging Bretz with their highest award. [9]

The Scablands geology doesn't necessarily prove the reality of the deluge of Noah's day. It does suggest the power of a surging wall of water as high as 800 feet, released by a ruptured ice dam, can cut a clean swath of jagged incisions in earth's crust within a few hours.

Since gradual river erosion or glacial-paced mountains of ice are not nature's exclusive tools for carving canyons, could it be that a worldwide hydraulic event supplied the ice age water that chiseled the Scablands?

When Mt. St. Helens blew its top in 1980, it cut deep gashes in the land while creating new strata layers within hours. The Scablands ice-water explosion cut its trail overnight. Neither earth-moving event took millions or even thousands of years.

Unlike the Scablands, a river runs through the heart of Arizona's Grand Canyon. Conventional theory posits that the most recent sediment layer at the surface took millions of years to accrue.

"The evolutionists' view is that a little bit of water eroded the Canyon over a long period of time through hard rock. The creationists' view is that a whole lot of water over a relatively short amount of time cut the Canyon through the still 'soft' rock layers laid down by the Flood." [10]

© Diana Lundin

Yosemite's Half-Dome, a silent sentinel to antiquity, marks the relentless march of time and the trail of a glacier that eroded its front half while tracing a trail through the valley.

The Bretz findings underscore the authority of water action in carving canyons and laying multiple sediment layers within brief moments of time. The Scablands scenario presents a persuasive case for hydraulic power scrambling a landscape within a few hours. Give that same surging force forty days and nights to do its thing, and the magnitude of the Genesis flood begins to come into focus.

"The average thickness of the sediments on all of the continents is approximately 1,500 meters...The average sedimentation rate measured over a period of one year is approximately 100 meters per thousand years." [11]

Is that rate of 100 meters per millennium reasonable? If relatively accurate, at this deposition rate all earth's sediment layers could have been put in place within 15,000 years, not a number compatible with the millions-of-years scenario.

Add a world-scale Flood to the mix and its further crunch time. The rate of sediment deposition would accelerate and that 15,000 year projected time frame would shrink radically.

The National Congress of Sedimentologists meeting in 1991 heard results that "...contradict the idea of the slow build up of one layer [of sediment] followed by another. The time scale is reduced from hundreds of millions of years to one or more cataclysms producing almost instantaneous laminae.

"These innocent-sounding words are the death knell of...the idea that the existence of thousands of meters of sediments is by itself evidence for a great age for the Earth...

Today, there are no known fossiliferous rocks forming anywhere in the world." [12]

A once pristine earth boasted a mild climate,
a carpet of lush forests, a plethora of jumbo-sized organic life forms,
and an atmosphere conducive to ecological balance.

Perfection reigned on Planet Earth. Every component of natural law thrived in synchronous harmony in a pulsating symphony of ecological balance. Complex formulas of atoms and molecules coalesced in a living, irreducibly complex ecosystem.

The Bible describes pre-flood human life extending several hundred years, even in deterioration mode. Unlike the commonly depicted, grunting, barely-human cave men, pre-flood humans possessed knowledge to invent, to explore the world, and the wisdom to record and to preserve written accounts of significant events and mind-expanding discovery.

Defective human choice disrupted perfection, dooming the landscape. Deterioration set in the moment the first human parents abused their freedom in ego-driven pursuit of faux "wisdom." Since that springtime of life, precipitous ecological decline followed, diminishing the natural world.

Eventually, nature's eco-skid was marked by the hydraulic cataclysm that Unleashed those "floodgates of heaven" [13] in an unrelenting downpour that raged, non-stop, for forty days and nights. When "...all the springs of the great deep burst forth" [13] to mix with 960 hours of heaven's open floodgates, the planet succumbed to devastating upheaval.

After forty days of pounding terror, "all the high mountains under the heavens were covered...to a depth of more than twenty feet." [13] The relentless inundation was intensified by "a wind over the earth," [14] unleashing Tsunami-like waves. Before the torrent of destruction was complete, hurricane force winds added to the raging fury, shredding earth's crust like a bolt of sheer cloth.

While much pre-flood knowledge would have been lost in the destructive waters of the deluge, the eight survivors would logically have preserved critical, encyclopedic information much as families today save heirlooms from a fire. Perhaps the remarkable shipbuilding skills, evident in the navies of the old Chinese Empire, came from Noah's ark experience.

If Noah and his family had used this narrow slice of time to record the drama of the most harrowing moments of their lives, preserved in written form, along with descriptions of the pre-flood world, the prized, museum quality records would have been passed to future generations. Moses, author of Genesis, would have had access to much of this data available through his royal education in Egypt and his unique cultural heritage.

More than that, Moses wrote under the Creator's direct inspiration.

The deluge event proved far more significant than the ruthless rampage of an angry nature; it exposed the sorry state of the human race. The exceptional longevity of pre-flood humankind only enhanced an already tainted environment where evil's contagion flourished.

While the "heroes of old" were "men of renown," they had become so prone to "violence" and degradation that "all people on earth had corrupted their ways" and "every inclination of the thoughts of his heart was only evil all the time." [15]

The Lord was "grieved...and His heart was filled with pain" when He "saw how great man's wickedness on the earth had become...the Lord said, 'I will wipe mankind, whom I have created, from the face of the earth.' " [15]

The meager traces of human beings in the fossil record is not due to some haphazard evolutionary process but, rather, to the catastrophic hydraulic event that not only buried coal beds and oil fields but also erased significant evidence of mankind's pre-flood existence. No wonder "the fossils that decorate our family tree are so scarce that there are still more scientists than specimens...all the physical evidence we have for human evolution can still be placed, with room to spare, inside a single coffin." [16]

Nothing remotely comparable to the global flood occurs today. The hydraulic action that scoured earth's face with catastrophic death and destruction was a one-time event. Still, after the days of Noah, nature's arsenal has demonstrated repeatedly its devastating power by unleashing fearsome destructive forces that are but something less than worldwide.

Localized floods, earthquakes, tectonic plate shifts, volcanic explosions, radical climate swings, the erosive power of hurricane winds, and the relentless grind of ocean tides have carved stone monuments. Tsunamis erase coastal villages, trigger fires resulting from natural gas explosions, and threaten all forms of life with terrorizing doses of radioactivity released by earthquake and water damage to vulnerable atomic power plants.

But never again will there be another global scale deluge.

"Whenever the rainbow appears in the clouds, I will wee it and remember the everlasting covenant between God and all living creatures of every kind on the earth." [17]

XIX

Tracking the Perfect Cataclysm

Fossil Residue

'Some kinds of catastrophic action is nearly always necessary for the burial and preservation of fossils. Nothing comparable to the tremendous fossiliferous beds of fish, mammals, reptiles...found in many places around the world is being formed today." [1]

John C. Whitcomb and Henry M. Morris

© Poutnik

Devil's Tower, remnant of a volcanic plug.

When organic life forms are suddenly swamped, buried, pressured by tons of sediment, deprived of oxygen, and subjected to heat, fossils can result.

A treasure chest of 20,000 fossils has been discovered buried fifty feet below the surface of a southwestern China mountain.

A team of "scientists led by Shixue Hu of the Chengdu Geological Center detailed their findings online Dec. 22 [2010] in the journal Proceedings of the Royal Society B." The cache includes "exceptionally well preserved" fossils "with more than half of them completely intact, including soft tissues." [2]

The presence of "fossil land plants suggest the marine community lived near a conifer forest" in a "tropical climate." [2]

The Permian mass extinction event, conventionally dated at 251.4 million years before the present, took out a majority of insects, an estimated 70% of terrestrial vertebrates and a staggering 96% of marine life. Even with sudden burial from catastrophic hydraulic action preventing early decay, "soft tissues" could not have survived for millions of years.

Extinctions were not solely the by-product of the failure of some ancient species to compete for survival and to transmit its "unaltered likeness" to some "distant futurity descendant."

Prime candidate cause for across-the-board extinctions of plant and animal life is the hydraulic powered event depicted in Genesis that tore apart an entire ecosystem. Dry land succumbed to an unprecedented water inundation wreaking environmental havoc. Species extinction and ecological decline remain the cataclysm's legacy.

Sudden inundation by water-borne sediment wiped out entire species, creating jumbled masses of disarticulated fossil bones, jammed together in fossil graveyards. This evidence correlates comfortably with the Genesis narrative of a relatively recent global gully-washer where water-powered devastation rearranged mountains of terrain, delivering indiscriminate collateral damage. Remnants of cataclysmic intrusion pervade dry land masses. Residual fossils provide clues to original biodiversity.

Another treasure-trove of tiny fossil embryos discovered in China were "...most likely buried alive one day in a sudden catastrophic overflow of sediment." [3]Absent such a process, dead organic material, exposed to the surface, will decay naturally and disappear in short time.

Marine life relics of pre-history have been discovered strewn across landmasses thousands of feet above sea level on the slopes of the Andes and the Himalayas, far from present day oceans.

"Fossilized dinosaur tracks scale sheer mountain cliffs, which are tilted topsy-turvy by some unseen, latent power—trademark testimony to the magnitude of cataclysmic force. Seashells and fossilized marine life litter bone-dry hilltops and mountain slopes, far above today's sea level.

"Chains of today's high-altitude, rugged terrain lay submerged underwater in the past, until sea beds awash in the currents of a cataclysmic deluge or powered by some convulsive thrust inside the Earth's crust pushed mountains skyward from the ocean's floor." [4]

The Siwaliks, foothills to the Himalayas, which run for several hundred miles and are 2,000 to 3,000 feet high "...contain extraordinarily rich beds crammed with fossils: hundreds of feet of sediment, packed with the jumbled bones of scores of extinct species...the remains of terrestrial animals, not marine creatures." [5]

"A fossil fish has been unearthed 17,000 feet up the slopes of the Andes and marine fossil limestone has been spotted in the Himalayas at an altitude of 20,000 feet!

"Marine fossils are found on top of glacial deposits as in the case of the whale skeletons...covering glacial deposits in Michigan...Whale fossils have also been found 440 feet above sea level north of Lake Ontario; more than 500 feet above sea level in Vermont; and some 600 feet above sea level in the Montreal area." [6]

The fossil bones of a whale were discovered on a desert hilltop near Bakersfield, California. Sea going creatures didn't simply crawl or swim up the mountains to die but are residual clues to a past natural disaster that gave Mother Nature a major league facelift.

With or without a villain meteorite killing off an entire species, it was sudden hydraulic action that fossilized Gobi dinosaurs. At the time it was a wetter and greener Gobi and home to "...hundreds of dinosaurs and mammals...Avalanches of water-soaked sand buried the animals alive, creating one of the world's richest fossil sites.

"The fossils appear remarkably complete in that...all the bones are connected to form whole skeletons...death was sudden, and the ill-fated creatures were quickly buried before scavenging animals could make off with the meaty bits." [7]

Thousands of fossil dinosaur eggs have been discovered, strewn across a parched square mile of layered mudstone within the Argentine badlands at the Auca Mahuevo site.

"Every evidence shows that the embryos may have perished in a flood that quickly buried the eggs in a layer of silt and mud. This made it possible for the soft tissues to fossilize before decaying, an extremely rare occurrence." [8] Irrespective of conjectured time frames, "dinosaur bones...had to fall into water and be buried to be preserved, and most dinosaurs spent most of their time on dry land." [9]

Nature's subsurface reservoir of carbon-based fuels is finite.
This biomass residue is not a naturally renewable resource.
Nowhere is nature manufacturing a replacement supply.

Mineshafts, honeycombing through thousands of feet of West Virginia's mountainous terrain, lead to a black biomass treasure that heats homes and generates electricity. Twenty-nine West Virginia miners lost their lives in April, 2010, while extracting coal from those underground tunnels that extend five-miles beneath the light of day.

Masses of compressed ferns and trees, a residue of another time, provide the raw material for those rich veins of coal several hundred feet deep. A once lush band of living vegetation, thriving under the sun's rays, didn't die a natural death. Coal didn't just bury itself but remains a visible footprint left by a colossal hydraulic event that moved and built mountains. Mother Earth underwent an unprecedented face-lift that shifted millions of tons of terra firma, crushing and creating the coal biomass.

Conventional time frames postulate a long and tedious process for forming the strata housing fossil fuels.

Is a multi-million-year process realistic?

Take another look at the mega-tons of rocky terrain piled thousands-of-feet deep atop coal beds. Conventional traditions centered on evolution's gradualism leave questions hanging, ignored and unanswered.

Where did this mountainous stack of stone originate?

How did it arrive in place to pile on and bury the biomass?

What source of power shuffled the mega loads of sediment?

Why isn't comparable action taking place today?

Fossil remains of trees have been discovered embedded upright in coal seams with vertical trunks pointing skyward, penetrating through multiple coal seam layers.

"Polystrate trees are fossil trees that extend through several layers of strata, often twenty feet or more in length. There is no doubt that this type of fossil was formed relatively quickly; otherwise it would have decomposed while waiting for strata to slowly accumulate around it." [10]

This polystrate tree phenomenon argues for sudden and near-simultaneous burial of multi-layered strata, posing a time quandary not resolved by multi-million-year scenarios.

The remains of plants and animals stranded on earth's surface, exposed to nature's ravages, will decay and disappear. But that same organic matter can provide the raw material for fossil fuels if buried suddenly, pressured by tons of sediment, deprived of oxygen, and subjected to heat. [11]

Peat bogs can't do it!

So when and how did this prodigious wealth of lush vegetation and animal life find itself buried in Mother Earth? A global-scale, catastrophic Flood stands out as the most reasonable explanation.

Commerce rides wheels greased by non-renewable biomass resources pumped 24/7 from hard-to-find hiding places. The world's largest supply of crude oil was discovered in 1938, 4,247 feet below the sands of the Saudi Arabian desert. [12] But once tapped and pumped dry, the pools of black gold treasure are gone for good. Civilization's voracious fuel consumption collides with sober reality. Production of oil and natural gas liquids peaked in 1970. [13]

Maryland's East/West bound Interstate #68 slices through a cross-section of breathtaking geology. Rocky Gap strata shelter a rainbow of earth tone colors presenting graceful arches of folded strata.

© Warren L. Johns

Despite some geologists' claim that hard rock can bend, it seems more likely that folded sediment would have been damp when bent into strata's colorful curves.

Enormous pressure must have shifted giant chunks of still damp sediments carving arced designs in the strata when solidified.

A raging hydraulic force, ravaging the face of Mother Earth in unison with other destructive events, could be labeled the "Perfect Cataclysm."

Earth's inorganic land and sea mass coalesce in a kaleidoscope of constant change, powered by the planet's internal heat engine, pushing to the surface, building a dynamic ecosystem stage.

© Kavram

Meandering ribbons of water, like this horseshoe bend in the Colorado River, display colorful canyon cuts preserving subtle clues to deep time.

Volcanic explosions, meteorite impacts, hurricane force winds, earthquakes, tectonic shifts, valleys chiseled by glaciers, shifting magnetic poles, drastic climate changes, radical extinctions of plants and animals, and rampaging water are not uncommon single events. The combo is unique.

If those natural forces were unleashed more or less simultaneously worldwide, as a package deal, the magnitude of destruction would be unprecedented. Something approximating that drastic package happened when "all the springs of the great deep" burst.

During the most recent 4,000 years of earth's history, examples of raging volcanic ovens, spewing mega tons of dust into the atmosphere, provide burning hot evidence demonstrating their ability to upset world climates.

Krakatoa exploded August 23, 1883 snuffing out 36,000 human lives while hurling ocean waves as high as fifty feet. The twenty billion cubic meters of ash and debris blown more than 20 miles into the stratosphere caused temperatures to drop precipitously by ten percent during the next three years, impacting climates as far distant as Europe. [14]

With a force estimated at thirty times that of Mount St. Helens, Alaska's Novarupta erupted on June 6, 1912, polluting the sky with a mantle of gaseous fumes and ash while lowering worldwide temperatures by as much as two degrees. [15]

In what is now the "Yellowstone caldera," the scene of at least three super volcano explosions in the past, the most severe is believed to have been "a thousand times the size of the Mount Helens eruption in 1980." [16]

"...Ocean temperature at the end of the Genesis Flood was likely as warm as 100° F or more. Such a warm ocean would be an explanation for the Ice Age because of the excessive evaporation of water into the atmosphere and deposition of snow in the Polar Regions and on mountaintops that would have occurred.

"An ocean with a SST [sea surface temperature] equal to or greater than 100°F would also likely have produced large frequencies and intensities of hurricanes beyond anything experienced today.

"...Giant hurricanes called hyper canes would likely have occurred over major portions of the earth. They would have grown to hundreds of miles in diameter, produced horizontal winds of over 300 miles per hour, had vertical winds of 100 miles per hour, and precipitated rain at rates greater than 10 inches per hour.

"Large amounts of erosion of the unconsolidated sediments would have occurred on the continents following the Flood. In this context, today's increasing hurricane activity represents a minor oscillation in the steady-state condition at the end of about 5,000 years of cooling." [17]

"...The abundant layers of lava and ash, mixed with sedimentary rocks around the world, attest to extensive volcanism during the flood...Requirements for an ice age are
a combination of much cooler summers and greater snowfall than in today's climate. ...Volcanic dust and aerosols remaining in the atmosphere following the Flood" [18]

The dust residue from Tambora's 1815 eruption upset climates 10,000 miles distant in New England, parts of Canada and Europe. The year 1816 was dubbed "the year without a summer," when an "unprecedented series of cold snaps chilled the area. Heavy snow fell in June, and frost caused crop failures in July and August. Sea ice was extensive in Hudson Bay and Davis Strait..." [19]

"What happened to bring on a cataclysm so widespread and abrupt?

"The answer smoldered 10,000 miles away, in today's Indonesia...the 13,000-foot volcano Tambora erupted on an island near Java. For a week thunderous explosions rocked the region and were heard a thousand miles away. Fiery ejections of rock, flame, gas, and steam shot into the stratosphere.

"Thirty-six cubic miles of earth blasted heavenward-the greatest release of energy ever known, dwarfing a nuclear explosion and even a nuclear war. At week's end 12,000 Javanese lay dead, tsunamis had killed thousands more on distant islands, and the volcano stood a mile shorter than before.

"The trillions of tons of material that Tambora shot into the atmosphere circled the earth with the winds. For more than a year they blocked sunlight from the northern hemisphere, dimming the planet with that 'sable hue." [20]

So what happened to excess water left from the Global Flood that devoured the earth? Did the wind sent to dry the Flood's waters contribute to the quick-freeze that introduced vast sheets of ice and an ice age, so bitterly frigid, that thousands of years later, parts of the planet are still clamped in its grip? Could the Flood's residue be the primary source of ice-age water or the glacier that buried Manhattan Island 300 feet deep?

The impact of volcanic ash on the global climate combined with the unprecedented magnitude of residual Flood water make prime suspects responsible for the steep temperature drop that activated a freezing climate.

Antarctica, Greenland, and the Arctic bear immense burdens of frozen water. One estimate suggests the ice and snow saturating the surfaces of Greenland and Antarctica together contain 70% of the planet's fresh water.

Snow masses still cap the poles and crown mountain ranges. Sea-bound glaciers creep slowly down craggy mountain slopes, feeding thirsty oceans that can sculpt shorelines.

"Perhaps as much as 95% of the ice near the poles could have accumulated in the first 500 years or so after the Flood...The 'annual' layers deep in the Greenland ice sheet may be related to individual storms rather than seasonal accumulations...

"Calculations of the number of layers laid down assuming the ice sheet accumulated rapidly near the bottom show that as many as 100 storms may have swept the polar regions each year accompanied by frequent volcanic eruptions." [21]

With ice layers towering three miles deep in some places and 90% of the earth's ice crowning Antarctica's land base, it's thought provoking that "three hundred miles (500 km) from the South Pole, sandstone beds lined with coal deposits...laid down in marshy conditions under a cool, moist climate" and that "fossil leaves and wood found in Antarctica indicate it was once warmer and forested." [22]

Greenland's ice sheet, "roughly the size of Mexico," covers 80% of its surface. In 2005, its "glaciers discharged more than twice as much ice as they did in 1996...enough fresh water to supply Los Angeles for 220 years." [23]

In the event the frozen water storage plants of Antarctica and Greenland should feel the heat of global warming and melt entirely, ocean levels could rise an estimated 230 feet.

Then there are tsunamis.

A balmy, after-Christmas Sunday morning, December 26, 2004, started like all previous holidays in Southeast Asia's favorite get-a-ways. Holiday celebrants basked beachside in the tropical breezes that bathed resorts lining the shores of the Indian Ocean. Until this moment, the "Tsunami" label existed outside the common vocabulary of most tourists.

But nature, sometimes a cruel teacher, delivered a crash course in demonstrating the unimaginable havoc possible from an undersea earthquake carrying a force equal to hundreds of atomic explosions. With little advance awareness, an angry ocean--generating waves surging at speeds up to 500 miles-per-hour--swept aside all in its path.

The grim reaper's scythe cut a swath through ocean shores in Tanzania, Kenya, Somalia, Seychelles, Maldives, India, Bangladesh, Burma, Sri Lanka, Indonesia, Thailand, and Malaysia. An all-consuming ocean, stoked by a Richter Scale 9.0 earthquake with an epicenter off the coast of Sumatra, snuffed out the lives of an estimated 175,000 victims. The impact shifted the geographic foundation of one Indonesian island.

A piece of the ocean floor more than 700 miles long (distance between Denver and Chicago) and 10 miles wide jolted 100 feet upward unleashing 135 cubic miles of churning water. The magnitude of the destruction removed entire towns and left battered human remains mixed with the debris of civilization strewn helter-skelter in grotesque heaps.

The ripple effect stretched ugly tentacles westward 3,750 miles to the east coast of Africa snuffing out the lives of 298 Somalians in the coastal village of Foar. The initial crushing onslaught ripped seashore lobster beds, depositing a harvest of death high up the slopes of adjacent hills. [24]

An angry Mother Nature wasn't through.

The tsunami nightmare swept Japan's shores March 11, 2011, triggered by another 9.0 Richter scale earthquake that shook the foundations of the nation. Shattering homes and shifting earth's axis, the power of this killer jolt pushed the nation's main island eight feet farther to the east. Showers of deadly radiation soon followed, released by ruptured components of vulnerable seacoast atomic power plants directly in the eye of the storm.

When the "fountains of the deep" exploded in the days of Noah, smashed tectonic plates would have broken land masses into continent-sized pieces of a ragged jig-saw puzzle, disrupting magnetic fields, scattering toxic clouds of climate-altering volcanic dust, and sending portions of earth's water blanket into deep-freeze mode.

Those same colliding tectonic plates, larded with fossil-bound remains of marine creatures, also could propel jagged mountain ranges skyward where oxygen was thin and temperatures cold.

Consider the disastrous effect of combining the Flood's roaring hydraulic turbulence, storm surges, rogue waves with the shake, rattle, and roll of multiple, magnitude nine plus earthquakes.

Then, blend in multiples of Indian and Pacific Ocean style tsunamis.

Finally, add a gargantuan cosmic discordance with a shower of meteorites cutting 120-mile wide craters as happened once in the seabed off Mexico's Yucatan Peninsula. [25]

Multiple natural disasters have scarred earth's face.
Still, no single destructive event compares to the ferocious cataclysm that swept life away more than 4,000 years before the present.

Ironically, some who accept the feasibility of earth's entire dinosaur population being destroyed by a meteor exploding off the Yucatan reject the possibility of a Global Flood that sabotaged primal perfection, diminished eco diversity, extinguished a broad spectrum of plant and animal life while scrambling land surfaces, leaving them unrecognizable.

Darwin (1809-1872) didn't live contemporaneously with Noah; nor did he offer convincing evidence supporting his debatable assertion that "no cataclysm has desolated the whole world." His unsupported personal bias is pitted against a vivid and detailed account of a Global Flood disaster written several thousand years nearer to the event than when Darwin ventured his opinion. His rush to judgment, denying a worldwide flood, leaves hanging a batch of questions.

Can giant-sized dinosaurs form fossils if covered gradually, at a minuscule rate, when preservation requires prompt and complete burial with multiple tons of sediment to avoid decay?

How can mass burials that created fossil cemeteries be explained apart from the churning force of hydraulic energy?

How could billions of barrels of petroleum and mega tons of coal form without submersion of flora, overwhelmed by the sudden surge of sediment-laden water?

What better rationale, other than hydraulic catastrophe, accounts for the preservation of disarticulated fossil remains of ancient plant and animal life and the unrenewable energy resources of coal and oil reserves buried deep within the heart of the earth?

XX

Man Hunt

Human Genealogy

"Is its explanatory power any more than verbal?
...Evolution not only conveys no knowledge, but seems somehow to
convey anti-knowledge, apparent knowledge which is
actually harmful to systematics..." [1]
Colin Patterson

**"Jesus loves the little children, All the children of the world,
Red and yellow, black and white,
All are precious in His sight,
Jesus loves the little children of the world."**

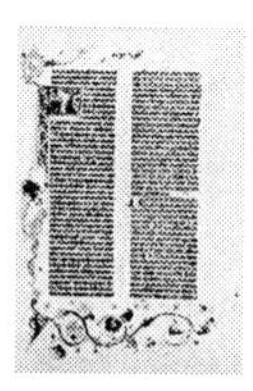

"To take a line of fossils and claim that they represent a lineage is not scientific hypothesis that can be tested, but an assertion that carries the same validity as a bedtime story, amusing, perhaps even instructive, but not scientific." [2]

On May 19, 2009, the fossil "Ida" was introduced to the world as "the scientific equivalent of the Holy Grail." [3] Split from a block of shale by amateurs in 1983, the lemur-like Ida measured a mere 50 or so centimeters from head-to-tip-of-tale

Under the auspices of New York's prestigious American Museum of Natural History, one enthusiastic sponsor glowed that the Ida fossil "will probably be the one that will be pictured in all textbooks for the next 100 years." [3] Another optimist declared Ida to be "the eighth wonder of the world," noting "we're not dealing with our grand-grand-grandmother, but perhaps with our grand, grand, grandaunt." [3]

A Princeton paleontologist discounted skeptics by welcoming the alleged 47-million-year-old fossil as "almost certainly part of the lineage that led to monkey, apes and humans." [4]

Still, Ida's auspicious introduction "as the eighth wonder of the world" didn't last long. Her fifteen-minutes of fame passed into oblivion abruptly, making way for "Ardi," another female fossil, destined to steal Ida's crown. Unearthed in 1994, "it took a multidisciplinary team 15 years to excavate Ardi, digitally remove distortions, and analyze her bones." [5]

Ardi's resume included the claim she lived 4.4 million years before the present, one million years before the date previously assigned to the still famous Lucy. One-hundred-twenty-five residue scraps of Ardi's fossil bones were pieced together with the help of computer technology. The composite result resembled a four-foot tall female ape, with "a body and brain only slightly larger than a chimpanzee." [5]

Capable of walking on two legs, Ardi's feet displayed a long, oblique-angled, "opposable big toe," [5] facilitating tree navigation. It took scientists mere months to conclude "the 47-million-year-old fossil of the early primate called Ida…is not a direct ancestor to humans, as initially claimed during her debut this year." [5] As to Ardi, "not all paleoanthropologists are convinced that *Ar. ramidus* was our ancestor or even a *hominin*." [5]

Does believed ability to stand erect necessarily equate "biped" status? Or qualify for sharing common ancestry with humans?

Evolution once relied on primitive paleontology's fossil bone fragments to conjecture linkage. Bone cemeteries can't do it.

Microbiology science looks beyond bones to DNA. Without clear-cut evidence of a long series of mutated DNA strings to corroborate evolution's postulated transitions, positive proof of Lucy's, Ida's or Ardi's imagined link to humans is wishful thinking.

Darwin never met Ardi, Lucy or Ida.
He advised humans should not "feel ashamed"
of common descent from "Old World division" monkeys.

After devising a theory built on chance, Darwin refused to acknowledge design in nature, while privately admitting to Harvard professor Asa Gray that he found himself " in an utterly hopeless muddle." [6]

In a culture proud of its ancestry and revered heraldry, evolution's champion diminished his own family's roots by claiming linkage to a long organic chain of life beginning with that accidental appearance of a first living cell derived from non-living matter.

In an in-your-face dogma of contempt, and without benefit of persuasive evidence, this godless theory mocks the creation miracle, alleging *Homo sapiens* represent nothing more than mutant descendants from an ancient fish, eventually sharing ancestry with "Old world division" monkeys.

The ancestral chain envisioned by Darwin linking man-to-molecule, by accidental luck-of-the-draw, defies reason. Such abstract linkage arouses skepticism, particularly so when incorporating a fantasy "fish story."

The philosopher/naturalist jump-starts the make-believe process by blazing a non-existent, transitional trail of organic continuity "through a long line of diversified forms" beginning with "higher mammals ...derived from an ancient marsupial " then latching onto "some reptile-like or some amphibian-like creature." [7]

With imagination unchecked, Darwin linked human genealogy to an unknown, pre-historic sea creature, asserting "all the members of the vertebrate kingdom are derived from some fish-like animal." [8] In his opinion, "Man is the co-descendant with other mammals of a common progenitor." [9]

Evolution's patron saint plunged ahead on a genealogical roll, saddling his own and mankind's family tree with yet another mysterious ancestor, envisioned as neither male nor female. "The early progenitor of all the Vertebrata must have been an aquatic animal, provided with branchiae, with the two sexes united in the same individual...." [10]

"Some extremely remote progenitor of the whole vertebrate kingdom appears to have been hermaphrodite or androgynous." [11]

He assumed "the progenitors of man must have been aquatic in their habits; for morphology plainly tells us that our lungs consist of a modified swim-bladder...the heart existed as a simple pulsating vessel." [11]

Darwin conjured-up man as having "descended from a hairy quadruped ...probably arboreal in its habits." [12]

Waxing eloquent, he insisted, "early progenitors of man were no doubt once covered with hair, both sexes having beards; their ears were pointed and capable of movement; and their bodies were provided with a tail." [13]

Stunning conjecture without a scintilla of solid supporting evidence. The linkage of successor life forms resulting from Darwin's postulated organic chain of physical traits, acquired by gradual use or disuse of body parts, over multi-millions of years, has yet to earn prominent feature recognition in human family photo albums.

The faces of Aunts "Lucy," "Ida" or "Ardi" are nowhere to be seen.

If Darwin's words are taken at face value, man's surmised ancestral linkage translates roughly: "fish-like animal...aquatic animal...reptile-like ...ancient marsupial...hairy quadruped...arboreal in its habits...sexes united in same individual... hermaphrodite or androgynous... covered with hair...both sexes having beards...ears were pointed, capable of movement ...higher mammals..."

Darwin floats this breath-taking rhetoric, despite the reality that this genealogical zoo of organic life forms--leading to and eventually linking "Old World division" monkeys with humans--doesn't exist. While not a highly visible admission printed in one of his published tomes, Darwin confessed in a personal letter to Asa Gray, "...I am quite conscious that my speculations run beyond the bounds of true science." [14]

He might well have added, far "beyond."

Born to a cocoon of wealth and high social status in a nineteenth-century British society with fiercely defined classes, Charles Darwin reflected a jaundiced view of life outside his circle of privilege. He wrapped his doctrine around a bias of narrow, social perspective.

He pictured himself and his male compatriots as mankind's fittest, surveying life from the peak of the heap, superior beneficiaries of the monkey-to-man scenario, leading the human race ever farther away from their alleged "arboreal" [12] roots.

Cross-sections of Darwin's unvarnished pronouncements raise twenty-first century eyebrows. Condescending prejudices infect his declarations. Consider samples of long discredited, politically incorrect thinking.

"Man has...become superior to woman." [15] "The average standard of mental power in man must be above that of a woman." [16] "The presence of...well-instructed men, who have not to labour for their daily bread, is important...as all high intellectual work is carried on by them, and on such work material progress of all kinds mainly depends." [17]"We must bear without complaining the undoubtedly bad effects of the weak surviving and propagating their kind." [18] "Races differ much from each other... the capacity of the lungs, the form and capacity of the skull ... in their intellectual, faculties." [19]

Darwin's imagined "tree of life," is at best,
an imaginary withered stick, bereft of leaves, branches or roots.

Darwin's "arboreal" assertion set the stage for the conjectured family tree descending from some grand pappy, monkey-type critter. Without a hint of equivocation, he attempted to pin-the-tale on the unsuspecting monkey for the entire human family.

He described the *Simiadae* as an "Old World division …after these had diverged from the New World division." [20] He disclosed the *Simiadae*s had "branched off into two great stems; the New World and Old World monkeys; and from the latter, at a remote period, Man, the wonder and glory of the Universe proceeded…" [22]

To minimize the shock value from this news impacting the considerable pride of peers, he encouraged followers to hold their heads high, assuring, "…we may, with our present knowledge, approximately recognize our heritage; nor need we feel ashamed of it…" [21]

Perhaps Charles wasn't ashamed of the monkey swinging from the Darwin family's woodwork, but his pointed portrayal of this presumed predecessor posing as some "arboreal" ancestor might not so easily have passed scrutiny when viewed through the discriminating lens of other Victorian families who took pride in the traditions of titles and heraldry.

Creationists may agree that humans deserve recognition as "the wonder and glory of the Universe," [21] while rejecting the dubious idea of life's accidental origin in some "little warm pond" followed by a tortured trail of transitional forms climbing relentlessly up the taxonomic chart to this prestigious pinnacle.

Has anyone heard even the most devout evolutionist boast of a scaly-skinned, Pisces ancestor, spawned in the briny deep or a hairy, ape-type ancestor noted for tree-swinging acrobatics?

This sorry fiction doesn't make the cut. Fact is, neither monkeys nor fish have anything to do with any human family tree, including Darwin's.

The same guy who saw no need to hang heads in embarrassment as to common bloodlines with ape-like ancestry betrayed personal blindness to racism, going out of his way to look down his nose figuratively on human races he labeled "savage." After leading readers all the way from primordial slime to the arboreal swing set, Darwin found a place for man parading from the "barbarian" or "savage state" onward to "lower races" and supposedly upward to "men of a superior class," [22] a position reserved to his perceived socially elite peers, dominating Europe's nineteenth-century.

Alexander Stephens, vice president of the short-lived Confederate States of America, couldn't have agreed more with racial categorization. In his 1861 "Cornerstone Speech," attempting to justify "the immediate cause of the late rupture and present revolution," the arrogant politician preached racial superiority as moral justification for slavery.

"Our new Government is founded upon…the great truth that the negro is not equal to the white man; that slavery, subordination to the superior race, is his natural and moral condition." [23] As a consequence, the American Civil War blood bath, fostered by blatant racism, snuffed out 620,000 lives.

Ever since Darwin conjectured man's divergence from " Old World division" monkeys, paleoanthropologists have scrambled, combing strata, looking for non-existent corroborating clues. Monkeys, bonobos and apes populate the African jungle scene but would be rarities in European or Middle Eastern landscapes. It comes as no surprise that Darwinists seek monkey/human common ancestry evidence where knuckle-dragging primate populations roam. Could the contemporary the "out of Africa" scenario represent an implicit corollary to the human/ape common ancestry theme?

Speculation suggests an out-of-Africa migration began 200,000 years before the present. This leaves a 150,000-year time gap until 50,000 years ago, when the Middle East appears to be the geographic hub of population migration[24] and the likely home base for most world languages.

Questions arise as to the validity of conventional time frames, as well as to the suggestion that early human migration took place out-of-Africa rather than "into Africa" from the Middle East. A potential blow to the out-of-Africa tradition is a recent claim that fossil evidence shows "anthropoid apes colonized Africa 39 million years ago rather than evolving there." [25]

Another snag in the out-of-Africa idea comes from a cave in China where "bones 60,000 years older than previous finds present 'a strong challenge' to the out-of-Africa theory and the 'traditional early-human time line.'" [26]

Given the human genetic traces emanating in most directions from the Middle East migration hub, a strong argument can be made that human migration routes led "into Africa" and "into Asia" rather than "out-of-Africa." After years of combing vast ranges of African turf, researchers have found pitifully few bits and pieces of fossil bones, even arguably categorized as pre-human. The clearly human Hofmeyr Skull, discovered in South Africa in the twentieth century, has been conventionally dated at 36,000 years before the present. [27]

But since conventional dating suggests migrations from the Middle East in multi-directions to all points on the compass occurred 50,000 years before the present, "Mr. Hofmeyr" might just as readily support "into Africa" theory as out-of-Africa thinking.

Before we swallow the allocated conventional dating of 36,000, 50,000 or 200,000 years before the present, the Hofmeyr Skull deserves a closer look. Mr. Hofmeyr's age was determined "by measuring the amount of radiation that had been absorbed by sand grains that filled the inside of the skull's braincase." [27] While initially this assessment may sound respectable, the troubling question lingers: Does the calculated date of Hofmeyer's burial necessarily match the date of the surrounding burial site turf? Without contrary evidence, could Hofmeyr arguably have walked the earth as recently as 4,000 or 5,000 years before the present?

As to Neanderthals, Bible scholars speculate as to whether the oversized *Nephilim* race of Genesis 6:4 and Numbers 13:33 could account for the Neanderthal fossils. However tempting the correlation, the speculative question rests on the table, unanswered.

Unable to identify human "evolution" in action in the living world, brazen phonies have tried to "salt the mine" with manufactured evidence, calculated to deceive.

Perhaps frustrated by the shortfall of evidence supporting evolution theory attempting to link human ancestry to unrelated animal species, blatant fraud took the stage. A weird parade of concocted "missing links" provides comic relief but does nothing to enhance Darwinian theory.

"Nebraska Man," purposed to be one million years old, grabbed 1922 headlines until exposed as nothing more than a single tooth of an extinct pig. *Ramapithecus* also missed the cut as an alleged "missing link" ancestor to *Homo sapiens* when ultimately identified as nothing more than an extinct orangutan. But the prize for arrogant deceit goes to the fabricators of the "Piltdown Man."

In 1912, Charles Dawson, a British lawyer and amateur fossil connoisseur, rose to the challenge, proclaiming his discovery of the fossil remains of "Piltdown Man." The fraudulent bone scrap claimed cache as an ancient fossil that tied apes and *Homo sapiens* to common ancestry. News of the find sent shivers of satisfaction down the spines of Darwin aficionados. The artfully darkened fossil skull and jaw shamelessly shouted "transitional," the long sought fruition of Darwin's unverifiable dream.

The Piltdown fabricator had pieced together a modern human skull with the jawbone of an ape, and then stained the monstrosity in order to mimic the appearance of antiquity. The faked evidence was then planted in an English gravel pit in prep for its orchestrated "discovery."

Evolutionists recognized as respected scientists promptly swallowed the bait. The ballyhooed fossil remains were promptly "authenticated" and awarded a place of honor in the British Museum. By the time Attorney Clarence Darrow took on the Scopes defense in Tennessee's 1925 "monkey trial," Piltdown's credentials reflected enshrinement in evolution's citadel of "fact." [28] In a written filing for the defense, University of Chicago Anthropologist Dr. Fay Cooper-Cole added his voice to the chorus preaching admiration for the find. Both Darrow and his expert witness had been hoodwinked, conned by the infectious ballyhoo touting egregious deceit. They bit on the phony discovery by cluttering the trial's record with written quotes classifying Piltdown Man as "...distinctly human...an approach toward man in very ancient strata...

"...The skull is exceedingly thick and its capacity much less than a modern man, but it is distinctly human, while, as indicated, the jaw approaches that of an anthropoid. Here again we seem to have an approach toward man in very ancient strata." [29]

The celebrated defense counsel further buttressed his shrewd defense by introducing citations from the 1914 edition of George W. Hunter's *Civic Biology*, the textbook then in use in Dayton, Tennessee's science classrooms.

The text overflowed with patently racist, survival of the fittest mentality, the kind of virus that infected Hitler's Germany in the 1930s. Ugly, unscientific garbage glared ominously from pages 195-6 of the despicably flawed volume.

"There is a greater difference between the lowest type of monkey and the highest type of ape than there is between the highest type of ape and the lowest savage...Undoubtedly there once lived upon the earth races of men who were much lower in their mental organization than the present inhabitants...we find that at first he must have been little better than one of the lower animals...

"At the present time there exist upon the earth five races or varieties of man, each different from the other in instincts, social customs, and, to an extent, in structure...the highest type of all, the Caucasians, represented by the civilized white inhabitants of Europe and America." [30]

Hunter's *Civic Biology* has long since been discarded, out-of-print, but not before corrupting young minds with toxic fraud.

Piltdown Man cheated "death" for an extended time, flummoxing evolutionists for nearly half the twentieth century. Mr. Piltdown thrived as scientific "gospel" for forty-one years. It took until November 21, 1953, for the much-hyped patchwork of old bones to be exposed as bald-faced fraud--the year DNA's double-helix design burst onto the stage, revolutionizing genomic science.

Evolution's curious quest to map a genetic trail leading from molecule-to-monkey-to-man remains lost in the jungle of ideas.

Mendel's law of genetics applies to all living organisms, including humans. Mendel's law and evolution theory are less than compatible.

The human body consists of an estimated fifty trillion cells composed of ten million atoms each. The DNA of a single cell reportedly comes packed with information sufficient to fill 3,000 encyclopedia sets. The human genome contains astounding variety potential, thanks to an estimated three billion DNA base pairs in the gene code. The original human couple carried a master mix of DNA enabling awesome variations among descendants.

This "random assortment of maternal and paternal sets of chromosomes at meiosis ...means that each human parent carries 8,388,608 gamete possibilities...A married couple has the possibility of producing over 70 trillion different children by this process alone (8,388,608 x 8,388,608)." [31]

There was a time when some scientists categorized human racial variances as different species, but no longer.

Today, all races are recognized as *Homo sapiens.*

Ever since time zero, when *Homo sapiens* first walked the planet, striking variety shows up in individual facial appearance, hair and skin color, eye expression and body size and shape.

John Hebert Johns (1889), Denver photographer.

Racial variances distinguishing earth's inhabitants result from migration and the isolated gene pools unique to a geographic area. Offspring of a couple representing any two isolated gene pools would only add to the diverse mix of humanity, never leading to some new and entirely different species.

Migrations to different parts of earth's geography, coupled with reliance on component portions of the pre-loaded master DNA gene pool possessed by migrating parents, guarantees descendant generation diversity. Distinctive markers include the mother's mitochondrial DNA and the father's "Y" chromosome passed along to sons.

Fully functioning human organisms blend millions of cells, tissues and organs into a single, synchronized whole, overwhelming evolution's dogma. For starters, the human body consists of an estimated fifty trillion cells, each composed of ten million atoms.

With more than seven billion humans walking today's earth, no person's identity is lost in the shuffle. Every individual owns a one-of-a-kind set of fingerprints and a unique molecular DNA code. The 23 human chromosome pairs in the cell's nucleus contain genetic information instructing each cell how to live and to reproduce.

The largest cell in the body is the female egg. The smallest is the male sperm. At the instant of conception, the human-to-be spends half an hour as a single cell. The tongue is the strongest body muscle. To make room for the heart, the left lung is smaller than the right.

The average human scalp has 100,000 hairs. Every square inch of the human body has an average of 32 million bacteria on it. [32]

It takes 200 muscles to take one step. Big toes have two bones each while the rest have three. A pair of human feet contains 250,000 sweat glands. About one trillion bacteria inhabit each foot.

Allegedly, the human body gives off enough heat in 30 minutes to bring half-a-gallon of water to a boil. Teeth start developing (in gums) 6 months before birth. The enamel in teeth is the hardest substance in the body. [33]

Humans come with a frame designed to walk upright on two legs without opposable big toes; a body needing clothing to match the weather; and language skills capable of conveying thought by sorting, recording and speaking thousands of words and phrases. [34] More than an organic automaton, the composite body system signals times to eat, sleep, and drink water. It also advises when thirst is slaked, hunger satisfied, and the system rested and refreshed for another day.

Uniquely human reactions trigger laughter, love, compassion, and loyalty, empowering pursuit of all things good.

Far more than a mechanical machine, a composite human has access to the same spiritual power that guided the lives of apostles Peter and Paul. Humans outrank all life forms on earth other than angels. The Bible narrative makes nearly 300 references to angels, supernatural created beings other than *Homo sapiens.*

Given dominion over all other life on earth, humans were honored to have been created in God's image, "a little lower than the angels." David, author of Psalms, reflects worshipful respect." What is man that thou art mindful of him?...Thou has made him a little lower than the angels." [35]

The apostle Paul confirms the existence of angels, informing they may appear among us in human form. "Be not forgetful to entertain strangers: for thereby some have entertained angels unawares." [36]

Population statistics and reproduction rates pose another, unresolved snag confronting the evolution model. Calculating a generation at 38 years with *Homo sapiens* reproducing for a million years, population estimates for today would be an off-the-wall impossibility even after factoring in "famine, disease, war, [and] natural calamity." [37]

"Studies have indicated the overwhelming majority of humans have a recent common ancestor within the last 5,000 years." "Identical ancestors point ...is estimated to be between 5,000 and 15,000 years ago." [38]

No evidence exists confirming humans share ancestry with "Old World division" monkeys as conjectured by Charles Darwin. Humans were created by God and have always lived *Sui generis* as a one-of-a-kind species.

Now and then overzealous headlines proclaim the début of a fossil alleged to be the common ancestor that humans share with chimps or apes.

Despite strenuous spinning, huffing and puffing, persuasive evidence that any simple, single cell fossil evolved upward from some "warm little pond" to *Homo sapiens* status continues missing.

XXI

Command Center

Human Brain

"Positive thoughts strengthen positive reaction chains
and release biochemicals, such as endorphins and serotonin,
from the brain's natural pharmacy. Bathed in these positive
environments, intellect flourishes, and with it,
mental and physical health." [1]

Carolyn Leaf

© Wes Kime

Physician Wesley Kime's oil portraits are reminiscent of John Singer Sargent's talent. Dr. Kime captured this vision of a colleague contemplating a life-saving protocol for a patient.

To claim the human brain evolved in random increments from non-living matter over mega-millions of years epitomizes the arrogance of ignorance.

Commonplace events that defy comprehension, much less scientific explanation, qualify as taken-for-granted "miracles." A man and a woman are vested with the power to share their genes and recreate a new human life. New humans arrive carrying the genetic code of two parents, blended in a one-of-a-kind mix.

No one expects a couple's union will produce some weird and different life form as forecast by evolution's deep time format. Still, some humans who swallow Darwin's "warm little pond" scenario as the kick-off point for human genealogy, expect their offspring to inherit traits carried by their own parental gene pool as long as harmful mutations don't intrude.

Misuse of the miraculous power to perpetuate life as nothing more than irresponsible recreation, prostituting a sacred gift. Original creation is ratified repeatedly by the privilege of parenting children, inspired by shared responsibility built on the mutuality of love.

Taken-for-granted sunbeams brighten the earth with miraculous, life-energizing rays powered by a super-heated furnace, all to a subdued chorus of ho-hum yawns.

And then there are those more than seven billion cerebral computers, each spinning off four billion calculations per second. The human brain generates more electrical impulses in a single day than all the world's telephones put together. The body package thrives with a mind loaded with creativity, capable of calculating complex equations and a freewheeling imagination able to concoct impossibly weird theories of life's beginning.

No rational person argues an electronic personal computer designed and assembled itself by random reactions of inert matter. Paradoxically, some intelligent minds insist human brains originated accidentally from a non-intelligent source, without design or designer.

Intelligent reasoning power from unintelligent matter? Talk about a leap of faith.

If human brains can concoct theories suggesting life evolved from non-life, why can't man's intelligence create a living cell that, like his first ancestor, created itself accidentally?

Gene expression profiles, residing in the brain's cerebral cortex, differ significantly between chimps and humans. Comparing the two, "…it seems that the brain is really special in that humans have accelerated patterns of gene activity." [2]

Give credit where credit is due.

No reputable university has awarded a Ph.D. to a chimp. Cute pets display clever behavior but never come close to an academic level that justifies a kindergarten graduation certificate. Chimps and dogs understand many human language words. Parrots and Lyrebirds mimic sounds and many spoken words and phrases. The fact remains, though: No animal brain matches the mind of *Homo sapiens*.

The human brain works creative wonders, thanks to a built-in pump and circulatory system. The heart and the brain function jointly as mutually dependent colleagues. Without a constant flow of blood to the brain, death arrives momentarily.

Brains orchestrate action in mega dimension beyond the capability of the most intricately designed electronic computer. A single brain *cell* is said to be able to hold five times as much information as contained in a set of printed Encyclopedia Britannica.

England's Stephen Wiltshire, a twenty-first century phenomenon, owns a brain that staggers imaginations of other mortals. Give Wiltshire a brief helicopter tour around London or Rome, and within a few short days he can draw what he saw on a curving, wall-sized stretch of paper. Doors, windows, roof designs, in right numbers and locations, an entire cityscape, all from memory.

Reasoning capability develops from a genetic code in place at conception, nine-months before birth. Thanks to DNA inherited from two parents, all creative thought, sensory reaction, speech, and memory are driven by "a three-pound brain…composed of twelve billion neurons…with…120 trillion connections" encased in a trauma-resistant, compact, cranial container. [3]

That memory bank "consists of about ten thousand million nerve cells.

"Each nerve cell puts out somewhere in the region of between ten thousand and one hundred thousand connecting fibres [fibers] by which it makes contact with other nerve cells in the brain. Altogether, the total number of connections in the human brain approaches 10^{15} or a thousand-million-million…

"Imagine an area about half the size of the USA (one million square miles) covered in a forest of trees containing ten thousand trees per square mile. If each tree contained one hundred thousand leaves the total number of leaves in the forest would be 10^{15}, equivalent to the number of connections in the human brain!" [4]

Consider the challenge confronting the world's most skilled electrician attempting to mastermind the wiring of that jungle of copper requiring a thousand-million-million connections without mis-wiring, short-circuiting, or blowing a fuse.

If that doesn't deflate the ignorance of egocentric arrogance, try confining that jumble to the micro dimensions of a three-pound brain fitted snugly within the custom-designed, bony cranium that typically offers something less than 100 cubic inches of spatial capacity.

So this feat of electrical/biological engineering supposedly designed and installed its own wiring diagram mechanism without so much as a master plan or Designer?

It's an irrational paradox to assert human intelligence derives its reasoning power from some unknown, unintelligent source. Humans, endowed with this incomparable "computer," are supposed to believe it evolved over many millions-of-years thanks to billions of trial-and-error, undirected, mindless, random mutations?

To assert human creative capacity originated from a non-creative, inferior, inanimate source ignores "rationality." When it comes to power to hang the moon, mere mortal genius lacks the "superior rationality" envisioned by rocket scientist Werner von Braun as the ultimate power "behind the existence of the universe." [4] The power source for gravity, magnetism, electricity and human thought derives from that "superior rationality."

Hong Kong neurosurgeon C. P. Yu sees the human brain as "the most complex 3 pound structure of this Universe." A Fellow of the Hong Kong College of Surgeons, Dr. Yu describes the "basic unit" consisting "of a neuron and glial cells" with at least "10 to 30 billion neurons and ten times that number of glial cells.

"Each neuron has 10,000 to 50,000 interconnections with other neurons. Electron microscopy differentiates excitatory from inhibitory neurons by the presence of a micro-spine." [5]

The nucleus of each neuron contains a "DNA molecule" which, if unraveled, would stretch a meter in length "within a cell 1/30,000 the size of a pinhead." [5]

Eight layers of bone with a "thick irregular plate" at the base "with openings for cranial nerves, blood vessels, and the spinal cord" compose a geometrically designed skull encasing the brain. Inside the skull, "the brain is surrounded by pressurized cerebrospinal fluid...providing nutrition and providing an active suspension system for the brain." [5]

Smooth movement of a coordinated body is taken for granted, until upset by the insidious Parkinson's disease, which disrupts basal ganglia function.

"...A tiny structure called the Subthalamic Nucleus (STN) is the pacemaker of the body. Numerous feedback loops and connections between the STN and other nuclei within the entire basal ganglia are responsible for the ultra-smoothness of movements." [5]

The brain orders the hand to jerk away instantly from a too hot surface, automatically, without conscious thought. The skin's power of touch equips the brain to respond to the hint of pain or to velvet's gentle caress.

Dr. Yu reminds that this message center for human life, manages five senses: smell, taste, hearing, touch, and vision. Feel, taste, sound, and smell impact thought processes that activate the full range of human emotions.

The brain "can distinguish more than 10,000 odors through tiny olfactory nerves at the roof of our nose." [5] And with that delicate sense of smell operational, taste buds introduce discriminating gourmets to the joys of dining.

Thanks to two ears with stereo capability, humans listen to and identify the direction of multi-directional sound. Three miniature bones, tiniest in the body, anchor the listening process.

Each ear comes equipped with "24,000 'hair cells', which convert vibrations to electrical impulses." The hearing nerve "enters the internal auditory meatus which houses 3 other nerves, 2 vestibular and 1 facial nerve, all tightly packed together and yet never pose any electrical leak or cross-over distortion." [5]

Look into the mirror and the evolution myth fades to irrelevancy. The human eye crowns a complex array of impeccably designed power!

According to Dr. Yu, "Apart from having auto-focus, auto-exposure, excellent low light response, excellent depth perception that no camera comes close, the eye can perceive: 1. Velocity, 2. Direction, 3. Location, 4. Texture, 5. Identity, and 6. Color." [5]

The neurosurgeon's summation exposes the preposterous notion of evolution's discordant dilemma, which attempts to account for the human brain's genesis. Screened through mathematical analysis of probability, complexity of this magnitude surpasses impossible.

"What is the probability of life arising from atoms to molecules, amino acids to protein (don't forget all life proteins are left-handed in configuration), DNA to messenger RNA, single cell to sexual reproduction, all the way to the human body with wonders of the brain and its senses, the heart and the circulation, the clotting cascade, the immune system, the wound repair and healing mechanisms?

"Bear in mind that all these have to go against the second law of thermodynamics, [the] law of irreducible complexity, and the fact that most mutations are harmful." [5]

The average human heart does its thing, beating rhythmically 100,800 times a day (with a pulse rate of 70 beats per minute) while pumping ten tons of blood daily, the weight equivalent of 140 adult humans. [6]

And here the plot thickens.

Already loaded with enough DNA in one microscopic cell to stretch from the earth to the moon and to dictate physical design, the brain is poised to absorb and record information from every sight, sound, smell and touch encountered in a lifetime.

Grc.NASA.gov

Intelligent human minds harness natural laws that open doors to space travel.

This acquired information is the raw material feeding the thought process that triggers responsive action. Anyone reacting to the flash of pain from touching a hot stove doesn't need the vote of a committee to instruct the hurting hand to move away from the danger. Thoughts produce action. When all is in sync, it doesn't take a genius to remove a hurting hand instantly from a hot stove.

Reaction time mimics lightning; some say the brain processes data as fast as four billion transactions per second.

There is danger that the cacophony of competing sounds and the blizzard of visual data confronting twenty-first century human brains can overwhelm its capacity to reason wisely. Despite the brain's remarkable information processing capability, anything more than seven competing items can compromise the brain's working memory.

"People faced with a plethora of choices are apt to make no decision at all…'A decision is harder if the amount of information you have to juggle is greater.' The proliferation of choices can create paralysis when the stakes are high and the information complex." [7]

All thoughts, whether good or evil, trigger attitudes and emotions. The composite thought process shapes character. An individual's ability to discover and accept the truth about God can be lost in the crescendo of information overload that attacks the senses.

"As he [a person] thinketh in his heart, so is he." [8]

"Positive attitudes cause the secretion of the correct amount of chemicals, and negative attitudes distort the chemical secretions in a way that disrupts their natural flow.

"The chemicals are like little cellular signals that translate the information of your thoughts into a physical reality in your body and mind, creating an emotion. The combination of thoughts, emotions and resulting attitudes, impacts your body in a positive or negative way." [9]

Patterns of toxic thoughts tend to put human immune systems at risk. A life drowning in evil thoughts, not only can resort to hate crimes and senseless murders, but also imposes a death sentence on the thinker as a long-term consequence.

"Buried feelings of anger, fear, anxiety and bitterness create volcanic buildups in your body. When you internalize wounded emotions, you allow a seething mix of anger, hostility and resentment to develop." [10]

The same Scripture that describes the miraculous origin of first life also offers the formula for abundant living.

"Get rid of all bitterness, rage and anger, brawling and slander, along with every form of malice. Be kind and compassionate to one another, forgiving each other, just as in Christ, God forgave you." [11]

The Creator pointed listeners to the merits of positive thinking that leads to the good life, assuring all mankind, "I have come that they may have life, and have it to the full." [12]

Science resonates precision and predictability. Toss a ball into the air and the law of gravity kicks in. The Periodic Table of the Elements provides a consistently reliable base for building chemical formulas. Ecological balance assures a life-friendly environment. Mendel's law of inheritance offers verifiable genetic results.

Surrounded by the bounties of a balanced bio-system, its logical to sense the presence and creative power of a "Superior Rationality."

Evolution's raw rejection of design put in place by a Master Designer reinterprets nature as merely a "Book of Random Accident," side-stepping confirmation of God's miraculous touch in the creation of life on earth.

Easily recognizable, pervasive evidence pointing to God's creative power has been twisted and overridden in an effort to convince gullible minds that life created itself accidentally in a "warm little pond" at some unknown time and place, millions of years ago.

Evolution responds with deafening silence when challenged to explain just how human intelligence, centered in the human brain, managed to manufacture itself and to acquire information from a non-living, unintelligent source.

Fence-straddling theists, swallowing evolution's bait, discount the authority of Scripture with the power and supremacy of the Lord, while clinging feebly to superficial "faith" in a manufactured, diminished divinity.

Darwin's tortured trail of flawed genealogy pushes mankind off the lofty pedestal of presiding leadership at earth's command center. Humans are downgraded to just another "biologic transit stop" moving aimlessly on a treadmill to oblivion.

Darwin fretted his grandiose scheme seemed "...a mere rag of an hypothesis with as many flaw[s] & holes as sound parts."[13] True science has yet to corroborate his wishful thinking. In fact, the "holes" he himself recognized have expanded to chasms and the series of "flaws" exposed as mere empty bubbles of irrelevancy.

In legal parlance, a cascade of exceptions "eats up the rule." Evolution's roster of "flaws" and "holes" devour any semblance of a reliable rule that may have been used to bolster and propagate a tarnished icon.

More than one-hundred-fifty years, and counting, after *Origin's* début, scientific knowledge has exploded but has been less than kind to a bankrupt concept, drowning in its own rhetoric.

The very definition of science, bent grotesquely out-of-shape by evolution's montage of assumptions, abstractions and suppositions, ends in a cul-de-sac of gross error. Evolution theory, in a variety of formats, has infiltrated human thinking insidiously for several thousand years. Heavy doses of hype and bias, have kept it in play.

Lacking rhyme, reason or respectability, the hypothesis manages to survive in the minds of imaginative humans, too proud to acknowledge the presence and oversight of a Higher Power, the Author of all science.

With a faint whiff of prescience, Charles Robert Darwin himself confessed doubts about his grand scheme, worrying that he may "have devoted my life to a phantasy." [14]

He had cause to worry!

Darwin's "phantasy" mirrored dreams of empire that rejected recognition of the Creator and categorized some humans as born to be subservient to others. Earlier, 56 stouthearted guys in Philadelphia turned their backs on fantasy, staking their lives and sacred honor on the premise that "all men are created equal."

Commitment to "unalienable rights," bestowed by God, the "Creator" of all life, anchored the legal framework for the birth of a nation. The operative words, "equal" and "created," echoed across the land each time the liberty bell chimed.

"The God who made the world and everything in it is the Lord of heaven and earth…He himself gives all men life and breath and everything else...From one man he made every nation of men, that they should inhabit the whole earth…" [15]

Darwin's evolution fiction never was, isn't now, and never will be.

XXII

Seven Billion Miracles

It Is About Us

"I find as difficult to understand a scientist who does not acknowledge the existence of a superior rationality behind the existence of the universe as it is to comprehend a theologian who would deny the advance of science." [1]

Werner von Braun

© Warren L. Johns

All human cultures share the *Homo sapiens* genetic code. Every person, like these bright-eyed Hong Kong youngsters, carries a one-of-a-kind genetic fingerprint

Good and evil co-exist, side-by-side.

When goodness triumphs, it makes headlines.

Its not everyday an unmarked envelope, loaded with $1200 cash, is lost on a public street in front of a fast-food restaurant, waiting for some curious pedestrian to discover. But that's what happened in downtown Murfreesboro, Tennessee, in March, 2010. The event made headlines because once the envelope was found, the unusual circumstances surrounding the discovery guaranteed widespread interest.

More than a year into the deepest economic downturn since the "Great Depression," and with unemployment reaching double digits, this kind of "mini-fortune" never grows on trees, -and rarely on city sidewalks. Typically, the finder might have pocketed the loose cash, reasoning "finders keepers, losers etc..." Fortunately for the owner/loser, the sharp-eyed young guy who spotted the windfall, proved to be anything but typical. Rather than depositing to his personal bank account, he entrusted the stash to the police in quest of the legitimate owner.

That's the upside of the story.

The downside soon followed as a surprising number of dishonest, greedy citizens reached out, pushing fictitious "claims" to steal the loot. When put to the test, the cheaters couldn't properly describe the envelope, the number and denominations of the bills much less the location in the city where the loss occurred.

The lucky owner, a truck driver, had no problem identifying his loss. Even before the police returned the $1200 to him, he met the good Samaritan finder and said "Thank you" with a $100 reward.

Goodness lives and thrives, even in the midst of a world overrun with the contagious virus of rampaging evil.

An aging Cherokee brave told his grandson of a battle that rages inside people. He said, "My son, the battle is between two wolves inside us all. One is Evil - It is anger, envy, jealousy, sorrow, regret, greed, arrogance, self-pity, guilt, resentment, inferiority, lies, false pride, superiority, and ego. The other is Good - It is joy, peace, love, hope, serenity, humility, kindness, benevolence, empathy, generosity, truth, compassion and faith."

The grandson thought about it and then asked his grandfather: "Which wolf wins?"

The wise old Cherokee replied knowingly, "The one you feed."

Legend's Diogenes traversed the world, carrying a lighted lantern, looking for an honest man. If he had lived in the computer age, he could have found him, right there in Murfreesboro, Tennessee.

Good and evil compete to control human minds and hearts, typically conflicting within the same person. Evolution theory offers no litmus test distinguishing right from wrong. *Malum en se* (wrong in itself) shifts and drifts; it is defined by the lowest common denominator in a prevalent culture. In a 1998 Darwin Day keynote address, one Darwinist admitted as much.

"No gods worth having exist; No life after death exists; No ultimate foundation for ethics exists; No ultimate meaning in life exists; Human free will is nonexistent." [2]

Evolution's dark philosophy---beginning in some in the warm pond and ending in certain death, is a slave to darkness. Evolution points only to indiscriminate darkness where "no ultimate foundation for ethics exists."

To extrapolate from the real to authenticate evolution's never-was, spreads an intellectual virus, infecting minds with toxic fraud. Preachers of pernicious, secular religion, seduce converts by calling the shots in a dark dance to nowhere.

Beyond philosophical speculation, evolution spawns a myth exposing humans to cultural darkness. With nothing but survival of the fittest in play, what evolves is anything but a pretty picture.

A furious Cain, in a fit of jealous rage, murdered his brother Abel, introducing Adam and Eve to the consequence of willful failure to trust God, the excruciating anguish of death. This black curtain of bitter hurt lingers, generation after generation.

Evil runs rampant. Before committing suicide in a Berlin bunker in 1945, Adolph Hitler had snuffed out millions of lives. Today, twenty-first century merchants of hate, misrepresenting themselves as God's agents, exemplify criminal cowardice, indiscriminately killing human beings they've never met.

Life's Creator epitomizes light.
The Genesis account of the miracle of life's origin
opens hearts to ten rules for better living.

When human ego succumbs to pride's seductive intoxication and evil infects the soul, the golden rule is tarnished, overwhelmed by an unconscionable greed that tramples property rights, degrading life.

There is reason to believe bully behavior delivers stress levels attacking the immune system while inducing an array of destructive heart and circulatory symptoms leading to disease.

Intolerance is symptomatic of insecurity. Evolution's intolerance of the truth about God and His creation sows seeds for a failed culture.

In contrast, unselfish caregivers, concerned for the health and well being of others while exhibiting low levels of aggression, tend to find peace of mind and longer life expectancies. [3]

Peace begins with a smile, and a clear conscience, marking a proven trail to abundant living.

Despite contrary myth, true science and true religion are not mutually exclusive. The two disciplines flourish compatibly. Reasoning that examines and evaluates all evidence objectively anchors and inspires both science and religion. Rational faith binds science and religion in an inextricably commingled package.

True religion is more than high-sounding theological rhetoric floating in fleecy clouds. Belief in the Creator of life is central to the understanding of life's origin, purpose, goal and ultimate destination.

Science is fact, and the Creator of the universe is its Author.

More than seven billion human miracles walk the earth.
Most free adults are invested with the intelligence to discover,
to design, to reason, and to discriminate right from wrong.

Though taken for granted, the birth of a baby is in reality a miracle.

Created in God's image, humans are more than an elaborate composite of rare physical prowess guided by minds capable of orchestrating music and launching rockets to the moon. The cohesive whole reaches out for a power that unleashes boundless dimensions of reason and creativity.

Apes, chimps and monkeys don't write books or sing songs. *Homo sapiens* do.

Human physical presence, powered by the mind's genius, has achieved astounding feats of creativity since ancient times. People explore space, produce movies, play golf, compose melodies, design instruments, orchestrate symphonies, build bridges, and invent machines as complex as a computer and as simple as a mousetrap.

Thanks to incredibly bright minds, replacement parts for autos can be manufactured using the NextEngine 3D scanner and Dimension 3D printer. Sounds like science-fiction but. as of 2011, it's reality.

Works of art from the hands, hearts and souls of Rembrandt, John Singer Sargent and Winslow Homer didn't appear on canvass as a result of a chimp swishing a brush dipped in a rainbow of oils. Leonardo DaVinci, Albert Einstein, and Galileo Galilei were born with mental acuity capable of sorting out heavy doses of information powering rare levels of creative genius. The phenomenal human brain possesses capacity for prodigious memory feats!

Whether myth or hyperbole, tradition reports Pliny the Elder, a Roman scholar, authored *Natural History*, a literary work suggesting Cyrus the Great, "knew the names of all the men in his army." Allegedly, Lucius Scipio was familiar with "the names of all the people of Rome" and "Mithridates of Pontus knew the languages of all the twenty-two peoples in his domains." [5]

Thomas Cranmer, Archbishop of Canterbury, is reputed to have memorized the entire Bible in three months. A blindfolded chess-master, George Koltanowski, played 56 matches simultaneously in a nine-hour marathon, winning fifty games while tying the other six. [5]

Then there's the occasional genius that astounds other high IQ elites.

Consider Wolfgang Amadeus Mozart, who walked the earth for a brief 35-years, composing volumes of musical scores that continue to enchant concert aficionados. The six-year-old musical genius made his first public virtuoso appearance in Linz. At ten, he performed a symphony of his own in Amsterdam.

During April, 1770, Amadeus visited Rome's Sistine Chapel where he listened to Allegri's *Miserere*. The Vatican reserved the complex musical to entertain honored guests in private concerts. The 14-year-old genius listened intently and later accessed his masterful memory to reproduce the entire score in flawless detail.

When honest intelligence combines with goodness, lives showcase the power of good trumping evil. The apostle Paul is a towering example.

Saul of Tarsus, a Roman citizen born of Hebrew parents, patrolled first century streets determined to root out the rapidly spreading Christian "heresy." In Saul's mind, half-measures wouldn't do. In an all-or-nothing context, the issue loomed as right or wrong, life or death. Belligerent intolerance gave no quarter. Once discovered by Saul's henchmen, any Christian faced a potential death sentence. Saul launched his anti-Christian hate crusade after standing-by, a witness to the stoning of Steven who met his fate, devoting his life to the "Righteous One."

Saul had seen Steven fall to his knees and heard him cry, " 'Lord do not hold this sin against them.' When he had said this, he fell asleep." [6]

Tasting blood, Saul went on a rampage, intent on destroying the fledgling church. "Going from house to house, he dragged off men and women and put them in prison." [7]

Zealous to a fault, Saul took his self-appointed task seriously, "breathing out murderous threats against the Lord's disciples." [8] He obtained letters of introduction to synagogues in Damascus, intending to find "any there who belonged to the Way, whether men or women," [9] determined to take them prisoner to Jerusalem, intending to do all he could to crush the Christian faith.

En route to Damascus, he did an abrupt about face. A blinding light sent him sprawling to the ground and an unseen voice instructed him to continue to Damascus where he would be told what to do.

The Creator of life on earth appeared personally to this brilliant man of honest conviction, both redirecting his mission and transforming his heart. In a flash, Saul, the persecutor, became Paul, the persecuted.

Due to his parents, the gutsy apostle possessed the privilege of Roman citizenship with ready access to its empire. He put these credentials to work in pursuit of his new mission. He sailed the Mediterranean three times, sharing the good news and founding Christian congregations. His time, talent and resources were devoted exclusively to all things good.

Paul the Apostle supported himself by making tents; he faced hostile audiences without fear; he presented the truth about God to political leaders; sometimes he ran for his life, narrowly escaping death; other times he went to prison for his bold presentations.

Ultimately he suffered a martyr's death, beheaded in Rome.

Paul's legacy of letters provided key components of the New Testament that preserved a clear picture of God's truth with its promise of life eternal, the reward of all people committed to the power and saving grace of Christ the Creator.

During the two thousand years since Paul's journey to Damascus was rerouted, a vast multitude of devout men and women have joined the international fellowship of believers, living lives of faith in action.

During World War II's raging fury, when sixty million human lives were snuffed out, many Christian men and women stood tall in the tradition of the apostle Paul, risking their own lives to save others!

Heroes like South Sea Islander chief Kato Ragoso, Cpl. Desmond Doss, and John Weidner put it all on the line for humanity's greater good.

Photo courtesy of Dr. Don Moran

Kato Ragoso risked his life to rescue struggling men from Pacific waters while WW II raged. He is shown here during a 1936 California visit.

Kato Ragoso, raised in a culture accustomed to vengeance killing, rejected a pagan heritage, embraced Christianity and dedicated his considerable leadership talents to telling the world the truth about God.

Visitors to a 1936 church convention in San Francisco were treated to the gentle voice and dignified demeanor of this Christian gentleman, a mere generation removed from a culture of violence.

Facing death each day in the midst of WW II hostilities, Kato organized his countrymen to participate courageously in life-saving missions. Without pay or fame, his team of unsung heroes patrolled Pacific waters, using a fleet of locally crafted canoes to rescue hundreds of struggling military men from capture or drowning.

Internet – My Hero

Cpl. Desmond T. Doss, unarmed World War II medic, was awarded the Congressional Medal of Honor for life-saving heroism, above and beyond the call of duty.

Future U.S. President, John F. Kennedy was saved by Ragoso's team.

By strange irony, this brilliant and courageous Solomon Island leader, who inspired hundreds of life-saving acts, would possibly have been viewed a "savage" if measured by Darwin's blind racism. It wasn't Western Europe's elite culture that miraculously transformed Ragoso's life, but the Christian faith that powered his soul.

When Desmond T. Doss, a humble Christian from the American south, served the United States Army as an unarmed medic during World War II, he took his faith with him, kneeling to pray bunk-side in Army barracks before retiring, ignoring the taunts. Once, while on his knees, an Army boot was cast his way by a scoffer.

All bullying derision disappeared once the 77th division launched into the Battle of Okinawa and encountered "a heavy concentration of artillery, mortar and machinegun fire" on "Hacksaw Ridge."

The only medic available to a 155-man company, the twice-wounded Doss, displayed "courage above and beyond the call of duty" by repeatedly exposing himself to the withering barrage sweeping the Maeda Escarpment and single-handedly rescuing more than 70 wounded soldiers.

This selfless man epitomized heroism and the essence of Christianity.

John Henry Weidner, another devout Christian, born in France of Dutch parents risked his life aiding more than 1,000 men and women to avoid capture by the Nazis. The famed "Dutch-Paris" underground was organized and fearlessly led by John and his sister, Gabrielle.

John, a master of disguise, was arrested by enemy occupiers more than once but always managed to escape. As for Gabrielle, the Nazis stormed a Paris church in the midst of a worship service, and unceremoniously whisked her away to a concentration camp. She didn't survive.

After World War II ended, France, Great Britain, Holland and the United States honored John's selfless courage by awarding medals of distinction.

The most persuasive sermons are lived, not preached from a pulpit.

Wikepedia

More than 1,000 men and women avoided capture by the Nazis during World War II, thanks to the "Dutch-Paris" underground organized by John Henry Weidner and his sister, Gabrielle.

Individuals adopt personal value systems ranging from aspirations for all things good to the depths of dark evil. Self-centered takers hurt and destroy what they touch, spreading doom and gloom. Christians return good for evil, reflecting the example of their Creator and the ten rules for more abundant, happier living outlined in Exodus 20.

More than mechanically articulating the truth about God, true believers invite the power that created the universe by "the word of the Lord," to enter their hearts and to guide their lives, embracing the same power source that created life and hung the moon.

Recognition of the reality of the creation miracle doesn't make you a Christian anymore than stepping into a garage makes you a car. Church pews house the faithful who walk-the-walk, alongside the mean spirited, hangers-on with exteriors professing a "form of godliness."

Evolution theory postulates natural selection, paired with mutations, can transform an organism into a physically different life kind, given mega chunks of transition time while Christianity promises that genuine human expression of faith in the Creator will transform a spiritually corrupt *Homo sapiens* into a "born again" person, assured of life eternal.

Recognizing humans have been created in God's image anchors understanding of life. Connecting to this power source, through faith based on evidence, opens the door to purposeful, "abundant living."

The roster of Christians having made this connection encompasses a multitude of dedicated heroes of faith, unsung by the media.

Insurance exec, "Stubby" Wall; clergyman Art Patzer; patient parochial school teacher, Helen Johnson; community leader, Clyde Unglesbee; widow ladies Bennett and Shepler; and three unsung churchmen surnamed Salisbury, Travis and Wheaton.

The list goes on!

Russia's Michael Kulukov; Singapore's Peter Foo; Spain's Daniel Basterra Montserrat; and flying missionary nurse, Dorothy Nelson; and Wayne Hubbs devotes time and resources to feeding the hungry and finding coats to shield underprivileged kids from the winter's chill.

This roll call of the devout would be incomplete without referencing Charlene Morrison Johns, selfless mother who gave me life, always doing what she could to make the world a better place.

These gentle Christians, many who have gone to their rest, along with unknown millions more, exemplify the power of faith in action, reaching out to God for the spiritual power to walk in His footsteps. All have left their own indelible footprints of love, honored in respectful memory!

Any person, walking with God, constitutes a majority, overwhelming darkness with the light of eternal life. "It takes just one solitary light to guide a thousand ships in from the night." [10]

Paul the Apostle said it best: "Since we are surrounded by such a great cloud of witnesses ...Let us run with perseverance the race marked out for us...and fix our eyes on Jesus, the author and perfecter of our faith." [11]

XXIII

The Truth About God

It's About Him

"You are not an accident...You were made by God and for God, and until you understand that, life will never make sense. Only in God do we discover our origin, our identity, our meaning, our purpose, our significance, and our destiny." [1]

Rick Warren

© cgd

An artist's perception of Christ the Redeemer Crowns Rio de Janeiro's Corcovado.

Choose your destiny--

life by random accident from rocks housing a "warm little pond,"

or life eternal, an unearned gift from the Rock of Ages.

A miracle is defined as "an event inexplicable by the laws of nature." [2]

Evolutionists and creationists alike must recognize the origin of first life on earth as miraculous, given the inclusiveness of that definition.

Devotees of the chance hypothesis conjecture life inexplicably created itself spontaneously from non-living matter, by the luck of the draw, millions of years before the present, absent design or designer. Creationists attribute the miraculous event to God's intelligent design, a gift from the living "Rock of Ages," not accidental generation, compliments of Mother Nature and Father Time.

Either view requires faith; the question is: Which brand of faith represents rationality?

The chance hypothesis belief in an abstract random series of accidents, in which life created itself from non-living matter, lacks the skimpiest credentials, offering instead a starvation diet of unsubstantiated myth.

Reliance on fortuitous convergence of inorganic matter by "natural" forces to explain life's genesis is, basically, a secular religion. It is worshiping at the feet of superstition, the antithesis of science, a bowing the knee to dark conjectures of the pagan occult.

Chance hypothesis envisions a thin soup recipe for life, a jungle menagerie of organic forms mindlessly competing for survival, destined for inevitable extinction. Hopelessly obsolete and irrelevant to our past, present, or future, evolution exudes toxicity. If it is correct, the world's future is past tense, bleak with the prognosis of forever death.

Attempts to wrap skimpy, imaginary surmising in the mantle of "science" sing from the same page as counterfeit religion. This shallow postulate articulates Asimov's "nothingness."

Suppose evolution's faux "science" represents actuality? Here and now, by random chance, gone tomorrow, without a trace or a legacy? Eternal death, the reward of evil, triumphs?

A sobering look at distorted mischaracterizations of the Creator plaguing the culture of Charles Darwin's era may partially explain his cynical take in his quest for a secular explanation for life's origin.

When Darwin lost his ten-year-old daughter Annie, to death, in 1851, the heart-wrenching blow squeezed joy from his sensitive soul. Europe's conflicted religious history offered meager solace for a father's aching heart. The influential naturalist had a front-row seat to observe pro-forma caretakers of faith "having a form of godliness but denying its power." [3]

Eight years after Annie's death, *The Origin of Species* was published with its jaundiced view of human existence.

Institutionalized religion, when subsidized and imposed on the public by the state's secular power, can propagate meaningless rhetoric diminishing and distorting the truth about God. Fraudulent caricature defames God as vindictive, inflicting pain on those daring to doubt.

Medieval fat-cat clergy often lived in comparative luxury while underprivileged citizens scrambled to survive. Crusaders, marching under religion's banner, drained wealth and destroyed lives in God's name.

The visionary, Galileo, lived under house arrest for disputing church tradition. His crime? Suggesting, correctly, the earth orbited the sun. Less fortunate dissenters faced torture on racks designed to disarticulate bones, inducing "confessions" from victims allegedly to "save" their souls. Some "heretics," refusing to recant, were burned at the stake.

Religious tyranny, furious at memory of John Wycliffe, the courageous scholar who translated the Bible into the English language to encourage public access, exhumed his body and burned his bones. Many medieval observers, repulsed by corrupt conduct masquerading as religion, found an excuse to reject religious pretense of any kind and to shrug-off God as an innocuous abstraction.

Honest believers risked being persecuted or viewed as weak and naïve.

Evil, posing as "good," misrepresents God. When counterfeit faith defames the truth about God, humans may believe they are rejecting God, when, in reality, they are turning their backs on a phony, disgraceful misrepresentation of God.

*****

Bright minds can power through mists of the unknown, searching for viable clues that might shed light on human existence and life's meaning.

In contrast to the abstract emptiness of the chance hypothesis, the Scriptural compilation of 66 books, written by more than 40 inspired authors over 1500 years, begins with Moses' Genesis narrative of the creation miracle and concludes with John the Revelator's description of Christ's return to earth bringing life eternal to faithful believers.

True religion is not an arbitrary, philosophical whim conjectured by mortal minds and propagated by tradition, but a miraculous inner revolution that conquers evil and moves lives toward the Creator.

More than meaningful religious insight or a record of the miraculous origin of life during a creation week of seven literal days, the Old and New Testament combine to present a "Big Picture" scenario that gives meaning and purpose to life, crowned with a future owned by mankind.

The Bible exposes intellectual dead ends while offering answers that bypass presumptive superficialities. The Scripture warns of "a way that seems right to a man, but in the end it leads to death." [4]

Overwhelming evolution's bleak philosophy, biblical faith worships the Author of science, the Supreme Being who created man in His own image and blazed the trail for victory over death. Science and religion coalesce comfortably as mutually compatible and supportive.

Looking to God is the beginning of all wisdom relating to life science. Recognition of true science leads to bowing knees reverently to its Author.

"For since the creation of the world God's invisible qualities---His eternal power and divine nature---have been clearly seen, being understood from what has been made, so that men are without excuse. For although they knew God, they neither glorified Him as God nor gave thanks to Him, but their thinking became futile and their foolish hearts were darkened. Although they claimed to be wise, they became fools...

"They exchanged the truth of God for a lie, and worshiped and served created things rather than the Creator..."[5]

"You alone are the Lord. You made the heavens, even the highest heavens, and all their starry hosts, the earth and all that is on it, the seas and all that is in them. You give life to everything..."[6]

*****

Scripture explains that evil originated in the heart of the angel Lucifer, the prestigious "light bearer" in God's presence. Vested with beauty and blessed with the power of choice, Lucifer's pride and jealousy led him to challenge God's authority, justice and love. Accusing God of tyranny, the insidious adversary, infatuated with his own malignant pride, stirred doubts, resentment and rebellion in the hearts of other angels.

"And there was war in heaven. Michael and his angels fought against the dragon, and the dragon and his angels fought back. But he was not strong enough, and they lost their place in heaven." [7]

Satan and his angels were banished from the presence of God and confined to earth, unleashing evil's fall-out of pain, death and devastation on an idyllic earth that once radiated ecological perfection.

"The great dragon was hurled down---that ancient serpent called the devil or Satan, who leads the whole world astray. He was hurled to the earth and his angels with him." [7]

When Satan was cast from God's heavenly presence, the newly created earth emerged as a cosmic theater featuring a spiritual battle performed before a universal audience. Played to its final act, the drama featured the tragic consequences of evil in contrast to God's love, fairness and justice.

The Creator could have suppressed the rebellion by destroying Lucifer and his followers; but had He done so, the act would appear to justify the charge of tyranny, with the result that other created beings, fearful for their own lives, would feel coerced to serve God from fear rather than from free choice inspired by love.

The Genesis narrative describes the cunning deception of Satan, the fallen Lucifer, now God's adversary, determined to mar a perfect creation and to recruit humans to join his rebellion. To showcase his power, Satan, all-time con artist and earth-bound adversary of all things good, focused his energy on convincing the first humans of God's alleged tyranny.

The father of lies set out to scar a pristine creation. Intent on revenge and expressing overt hatred for God, the adversary not only corrupted a perfect ecosystem but inspired the deceptive fraud that life created itself from non-life, abracadabra style.

No human reclines in a box seat as a bemused spectator.
Each plays a supporting role in the spiritual drama,
center stage in the "Theater of the Universe."

"We have been made a spectacle to the whole universe, to angels as well as to men." [8]

With Planet Earth occupying a miniscule blip on the cosmic landscape, human beings have been reminded of their pedestal status, created in God's image, and that not even a sparrow "is forgotten by God." [9]

Adam and Eve lived originally as free moral agents in an idyllic environment, sustained by access to the Garden of Eden's "Tree of Life." The "Tree of Knowledge" remained off-limits, simply to test their trust in and allegiance to their Creator.

Satan deceived Adam and Eve by appealing to their egos. He promised, by their tasting the fruit of the forbidden tree that their eyes would be opened and they would acquire the wisdom he alleged God had denied them unfairly.

The tower of lies came crashing down when the first couple succumbed to the fabrication that willful disobedience would not lead to death. Evicted from the Garden for failing the test, the pair did not die immediately, but the debilitating death process set in. The phony promised "wisdom" proved nothing more than exposure to the bitter knowledge of the consequences of evil.

Exposed to the killer virus infecting the soul, Adam and Eve lived to endure the agony of death in their immediate family when an enraged Cain murdered his younger brother Abel.

Denied face-to-face contact with the Creator's "superior rationality," [10] all creation suffered deterioration. Cut off from access to the "Tree of Life," humans could not escape death's curse, inevitably returning "to the ground since from it you were taken." [11]

The "Big Picture" perspective suggests mankind's culpability in the decline and fall from perfection of all creation. Bleak consequence of disobedience to God ravished the pristine environment.

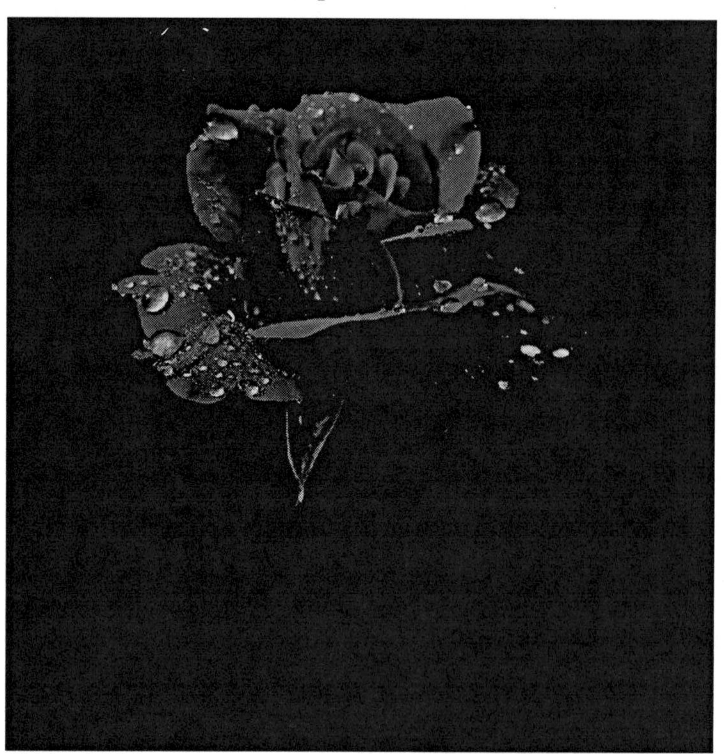

© Chuck Nelson

"Thorns and thistles" invaded once fertile soil.
"Cursed is the ground because of you.
It will produce thorns and thistles for you." [12]

The end-time spiritual picture includes cunning deceptions reminiscent of Satan's misleading promise of "knowledge" that enticed Adam and Eve.

Just as the first human couple were corrupted by blatant falsehoods--bright, sophisticated minds are vulnerable to the subtle seduction of "the secret power of the lawless one... already at work" and of the "powerful delusion" that advances "the work of Satan displayed in all kinds of counterfeit miracles, signs, and wonders and every sort of evil that deceives." [13]

Having deceived Adam and Eve in the Garden of Eden, the diabolical prevaricator initiated the same deceptive strategy to discredit God by depicting the Scriptural creation narrative as myth and life on earth as nothing more than a mindless, evolving accident rooted in a cacophony of chaos. Always the deceiver, Satan wraps his poison pill in the glorification of transient power.

Scripture warns Satan will appear "proclaiming himself to be God," [13] impersonating the Creator and counterfeiting Christ's return to earth. A dazzling figure, speaking pious platitudes, can be expected to deceive all but the most astute.

True religion bows humbly at the feet of the Lord God Almighty, the Creator of the universe and all human life. Unlike pagan superstition, abstract philosophies, and distorted faith, Christianity builds faith on a God who demonstrated His power, His love and His justice by assuming human identity, living a perfect life, voluntarily dying for the failures of the entire human race, and then conquering death for all who believe and follow.

He's alive!!!!

No other religious system compares!

The evil versus good drama reaches a showdown moment when the crucified Christ eventually returns to earth. "He is coming with the clouds and every eye will see him." [14]

Reminiscent of His own triumph over death, Christ will resurrect all who have died in the faith, award eternal life to all who have walked in his footsteps, make all things new, and destroy evil for all time.

XXIV

God's Greatest Gift

Life:---"Original, Unborrowed, Underived"

"King of Kings. Lord of Lords. He shall reign forever and ever." [1]

George Frederic Handel

© MaxFX

Life's origin on earth makes no sense without recognizing the Creator as the source of all life.

"It is finished"[2] were the last words heard from the anguished lips of a victim suffering excruciating pain, condemned to a death of cruel punishment by a tyrannical regime that had contempt for human life and equated justice with the power of the sword.

Events of that Friday afternoon, early in the first century A.D., changed history forever and played to a universal audience in the cosmic theater.

Face bleeding from a crown of thorns jammed onto his head, taunted with a purple robe mocking kingly power; brutal blows to the face, a public flogging--Christ endured all before being found guilty of anything.

His crime? Three years serving a suppressed populace with a message of healing, hope, and happiness. After questioning Jesus of Nazareth, the pathetic Pilate declared, "I find no basis for a charge against Him." [3]

Still, the unruly mob of vigilantes ruled courtyard chaos. In a night sham trial, Pilate, a weak-kneed political hack--intimidated by frenzied fanatics that thumbed collective noses at legal due process--sent an innocent thirty-three-year-old to His death.

Execution on a cross represented Roman "justice" at its worst. Christ surrendered His life on the cross, voluntarily. A jeering crowd trailed His steps as Roman soldiers placed a cross on His shoulders and led the way to Jerusalem's "Hill of a Skull," the execution site.

Torture was added to the mix by nailing the hands and feet to rough-cut lumber. The primitive device was lifted and then thrust into the ground to maximize hurt. While facing taunts from the heartless, the victim was left to writhe in pain for hours, or even days, until spared with a final gasp of breath, cruel punishment even for the most despicable criminal.

When Christ breathed his last, the sun refused to shine for three hours.

Minutes after His death, He was buried. No funeral or memorial service honored His life. Nor was He recognized with public burial rites featuring eloquent rhetoric. His death matched his birth's humble circumstances.

Joseph of Arimathaea volunteered his own family's tomb, chiseled from hillside rock, as a final resting place. But the "rest" was not to be "final." After the Sabbath, early the first day of the week, the seal of the tomb was broken and the stone entry door rolled away by unseen hands. Christ rose, forever victorious over evil and death.

Christ's victory marked a landmark moment, a turning point in world history. The resurrection miracle assured eternal life for all believers and sealed the death sentence on God's adversary. Christ's immaculately conceived birth, flawless life in the midst of an evil environment, cruel crucifixion and unselfish death, crowned by the resurrection miracle, is unmatched in other belief systems.

Satan tasted the sting of defeat in the raging controversy between good and evil. His eventual extinction would be but a matter of time. Christ the Creator, God's son in human form, lived and died to demonstrate to the universe that Lucifer's hatred would kill even God, if he could.

But the apparent "triumph" of evil proved ephemeral.

The Scriptural lexicon defines evil as sin that can "deceive even the elect." [4] Its ripple effect disfigures perfection and devours life, even God's son, a casualty of the spiritual war pitting good versus evil.

Despite imposition of death's curse imposed on mankind for disobedience and failure to trust, God "will provide a way out" [5] through the plan of redemption demonstrating His love, fairness and ultimate justice. God's "one and only Son," [6] whose perfect life and ministry demonstrated limitless love, took the hit and suffered the death penalty for all sinners.

Late in the fourth century, Aurelius Augustinus (Augustine), Bishop of Hippo, experienced and embraced the "Big Picture."

Christ "was righteous as God is righteous; and because the reward of righteousness is life and peace, He could, with His righteousness, united with God, cancel the death of justified sinners…That they might be saved through faith." [7]

Until Christ visited earth in person, the image of God's life-giving power and created perfection was blurred, lost in contrived trivia and a cosmic jumble. Christ's sacrifice revealed Satan's raging hatred and God's unbounded love for all choosing to believe and accept His promises. God's Spirit, with His righteous power for forgiveness, justifies the life of any believer.

*****

Jesus, of Nazareth, the greatest life ever lived on Planet Earth, arrived with less than auspicious human credentials. He was born in a barn, burdened by rumors of suspect parentage. Less than a bed-of-roses environment, corruption shrouded the streets and alleys of his home village. Surrounded by poverty, Jesus walked tall in a culture struggling for survival under the boot of Rome's iron empire. Life was cheap and the blood of gladiators drenched the sands of stadiums in order to satiate a depraved, public thirst.

Christ served an apprenticeship in Joseph's Nazareth carpenter shop. Nathaniel, a skeptical future disciple, aware of the notoriety of Christ's hometown, asked rhetorically, "Can any good thing come from there?" [8]

Christ never had a bank account, owned a home, wrote a book, held political office or commanded a military force.

Exuding charismatic personal charm, all it took were the words, "Follow me" for twelve young men to join his unique ministry of hope, goodness and the promise of abundant living today and a forever life to come. His disciples followed with no promise of pay or explanation of just where He would lead. No one asked about perks; they accepted his invitation simply because He said to them, "Come, follow me... and I will make you fishers of men." [9]

Some misunderstood the invitation, hoping that the Messiah would deliver subjugated citizens from the heel of the Roman oppressor. Despite lack of academic credentials, He taught thousands, captivating minds with a message of love, forgiveness, and a formula for peace. He lived and led by example from the geographic crossroads of world commerce. Without access to mass media, news of His words and deeds traveled with lightning speed. No media blitz publicized His appearances. Rumors that Jesus worked miracles, attracted attention to His ministry.

He didn't sermonize from the altar of a grand cathedral or a pulpit in an arena seating thousands. Still, multitudes followed His footsteps, captivated by a message that echoed across cultures.

When a wedding celebration ran short of wine, He transformed water into prime vintage by His word alone. Powered by His blessing, five loaves of bread and two fish satisfied the hunger of a crowd of 5,000. With leprosy's blight running rampant, hearts beat faster and eyebrows raised with news He restored vibrant health to ten lepers. Responding to His command, an infirm man "picked up his mat and walked." [10] The restoration of sight to a blind man startled witnesses to the unprecedented scene.

The miracle of bringing Lazarus back to life, days after death, defied all finite norms. This evidence of power over life itself underscored Christ's authority to create.

People in the countryside pondered His wisdom and responded to His leadership. Looking for clues to empowerment, hundreds followed Him to a natural, outdoor setting on the slope of a hillside. Admission tickets were not required. He spoke with authority.

But to the surprise and chagrin of some who came prepared to take up arms against the Roman occupier, He offered words designed to galvanize the human spirit, sounding a call to war against evil, not to military action.

Contrary to epic notions of heroic history built around military conquest with its harvest of death and destruction, Christ created original life and preached the sanctity of life with peace and goodwill among men. Against all odds, he predicted the "meek" would "inherit the earth." His bold advocacy revolutionized conventional traditions.

"Blessed are the merciful, for they shall be shown mercy. Blessed are the peacemakers, for they shall be called the children of God…." [11] He urged listeners to seek right, to be pure in heart, to love enemies, to forgive up to seventy-times-seven, and to reconcile with adversaries.

As the antithesis of darkness, He reminded followers: "You are the light of the world…Let your light shine before men that they may see your good deeds and praise your Father in heaven." [12]

© An Alan Collins sculpture, Loma Linda University Medical Center

"Who touched me?"

This formula for a better life opened the door to personal peace and abundant living. It came with the gift of a Divine Comforter, empowering believers to grow into a new and improved version of themselves.

The crucifixion events, crowned by the resurrection miracle, described by eyewitnesses in the writings of Mathew, Mark, Luke and John, are preserved for all to read in the biblical cannon, composed of Old and New Testaments compiled by Christian leaders late in the fourth century A.D.

Its an irrational paradox that some, who doubt Christ's resurrection and the Genesis narrative of the creation of first life on Planet Earth, swear allegiance to an unsubstantiated chance hypothesis that life on earth might have created itself spontaneously, from non-living matter, independent of intelligent design or Designer.

Christ exemplified God's absolute justice and patient love for humanity, exhibiting unlimited Divine power, not just to hang the sun and the moon but also to conquer death's curse and to exterminate all taints of evil and its culture of death. His selfless ministry revolutionized history.

Tyrannical rulers don't hesitate killing their own subjects in order to preserve power. In contrast, the ruler of the universe and Creator of all life, not only voluntarily gave His life to demonstrate the power of love, but did so for the benefit of humans who had joined Satan's rebellion against Him.

"Greater love has no man than this, that one lay down his life for his friends. You are my friends if you do what I command. This is my command: Love each other." [13]

Christ would have willingly suffered death if for only one person, even an enemy. His unassailable victory over sin and death, as reported and recorded by living witnesses, left God's adversary little option but to surrender to failure or to attempt to rewrite history by frontal attack.

Although Christ committed no crime, His human body died the death of a criminal, an innocent victim of man's inhumanity to man. While suffering excruciating pain, Christ forgave his tormentors with the words, "Father forgiver them for they know not what they do." [14]

Forgiving enemies who have "trespassed against us" may appear to be a bitter pill to swallow---but it is key to the cure cleansing the inner person. The Bible promises: "If we confess our sins, he is faithful and just and will forgive us our sins and purify us from all unrighteousness." [15]

God forgives all sins and any sinner who asks forgiveness.[7] The truly forgiven follow Christ's example---the exclusive route to inner peace. Without forgiveness, festering hatreds spawn war with its death curse.

Peace, in contrast, begins with forgiveness, and with a smile.

A wise twentieth century observer noted: "It is easy enough to be friendly to one's friends. But to befriend the one who regards himself as your enemy is the quintessence of true religion." [16]

Injustice, assassinations and wars result when anger, bitterness and hate are projected into a family, with ripples infecting communities and nations. Hatred generates stress, corroding the soul. Dr. Caroline Leaf goes so far to suggest the possibility that "your thoughts create changes right down to genetic levels, restructuring the cell's makeup." [17]

A personal living faith inspires forgiveness, extending unconditionally to enemies who despitefully use and abuse.

"Forgiveness is a choice…there is increasing scientific evidence that forgiveness gives us healthier and happier lives…When people hold onto their anger and past trauma, the stress response stays active, making them sick mentally and physically." [18]

It may not sound like a bed of roses to return good for evil and to reach out in forgive no less than "seventy time seven," but its feasible when the power to forgive comes directly from the Creator of light and all life.

Contrasting worldviews offer each human an either/or choice.
The evolution/creation debate reaches beyond
academic sparring over dueling dogmas.

The Bible's Joshua challenged followers to discard "the gods your forefathers worshipped beyond the River and in Egypt, and serve the Lord…Choose for yourselves this day whom you will serve…As for me and my household, we will serve the Lord." [19]

Humans possess the power to reason and to choose an eternal destiny. One view postulates evolution's random chance odds. The other embraces design with meaning, created by the Author of science. Nothingness exists in between but unplanned accident, an intellectual no-man's land.

Every life plan without God, at the core of the equation, suffers from terminal limitation as to purpose, scope and longevity. It's a life or death decision. Moral choice is not without cost.

False pride feeds selfish egos, corrupting hearts and empowering sin. "For the wages of sin is death, but the gift of God is eternal life in Christ Jesus our Lord." [20]

God points to a new and drastically improved life style. Our consciously choosing, by faith, to connect to infinite power opens the door to abundant living's rewarding adventure.

"The Lord is gracious and compassionate, slow to anger and rich in love. The Lord is good to all; he has compassion on all He has made." [21]

God's power to create original life includes the power to redirect a depraved life from degradation toward unselfish goodness, transformation of a wicked person into a "born again" spiritual being. Beyond cosmetic superficiality, the inner person enjoys a miraculous recreation reflecting Christ's image in thought and deed, accompanied by an unsurpassed peace.

The most influential sermons are not preached from a pulpit but are lived on the street, in the office and in the home. Talk-the-talk verbiage lacks the impact, influence and result of walk-the-walk action.

Men and women of faith, who have walked the earth from the beginning of time from all eras of history, comprise a royal line of faithful believers. Something more than charismatic attraction inspired a royal line of believers--like Noah, Abraham, Daniel, Moses, Peter, John and the blessed Polycarp--to put their lives on the line in commitment to the Creator.

Evolution is a fact, but not the transition of an organism to an entirely new and different life format but rather the miraculous transformation of a degraded human life into a born again person, growing in grace.

The crucified thief, enduring his final moments of life on a cross next to Christ, by faith accepted the gift of Christ's perfection, blotting out his dismal record of evil in exchange for the promise of life eternal.

God exposed Saul of Tarsus to the same light that brightened Planet Earth on the first day of creation week, powering a revolution in his heart. Saul, the persecutor embraced that vision, voluntarily taking on his new, spiritually driven identity as Paul the Apostle.

A battle between good and evil rages within each of us; it's a spinoff from the cosmic spiritual war. Evil's toxic and degrading disease infects and destroys. Good triumphs when God is invited to direct a human life. Any person, allied with the Creator, constitutes a majority.

Knowledge alone, without commitment to spiritual renewal, lacks substance. Intellectual comprehension of the truth about God and the creation miracle is worthy but is worthless and self-deluding without accepting Christ's promise to recreate and renew your inner person. Access to the power that created all things is available to everyone, conditioned only on the individual's voluntary free choice built on trust and expressed in humble rhetoric.

"Create in me a pure heart, O God, and renew a steadfast spirit within me." [22]

Getting acquainted with God and learning to walk in His footsteps involves a spiritually maturing lifelong experience. It is never too early, or too late, to select the role you choose in the big picture now playing in the cosmic theater. God provides the power. The choice is personal, unrelated to any third party or organization.

Biology and environmental factors influence each inner person battle.

"Biology clearly has a strong effect on social tendencies--a theory that is as strongly supported as anything in science...While genetics and even epigenetic inheritance certainly affects our social tendencies and natural inclinations, God has also provided us with access to supernatural powers that can be called upon to elevate us above these inherited tendencies...

"Social behavior is not absolutely tied to biology. This is why we still have access to free will...When one asks God for power to overcome this or that natural or cultivated tendency, and that prayer is dramatically answered, the power that is provided to resist what once was an overpowering urge to act in a detrimental way is strong evidence of the existence of a Power that is outside of and greater than one's self." [23]

Evil won't relinquish its grip on the strongest willed person that fails to choose to connect to the infinite power who built the universe and created life. Once human faith accepts the truth about God, life reverses course and moves upscale. "The peace of God, which passeth all understanding," [24] buoys hearts like a golden cloud floating on cool sunshine.

Regardless of a blemished lifetime of wallowing in evil's muck and mire, it is never too late to invite the Creator to redirect the inner person. The Creator can recreate any human heart. No person lives beyond the reach of God's love. Each day offers a blank page with doors open wide to the abundant living that accompanies life-changing miracles. Walking in the Lord's footsteps, degraded characters scarred with pride, hate, cunning and greed are rebuilt by the Creator's empowering forgiveness and love.

When you choose fellowship with Christ--doubt, discouragement and depression can be banished. Sin-marred pasts are obliterated. Reorientation with radical character modification takes over, any time a human heart, saturated in evil, surrenders body and soul, embracing God's power.

Children depend on parents for survival. Christ counseled, "Unless you change and become like little children, you will never enter the kingdom of heaven." [25]

Character building involves a lifetime commitment. The occasional good deed or misdeed is not the final test. Humans are judged by the direction taken, not by occasional stumbles.

Instinctively, and without hesitation, men and women rush into water to save the life of a drowning child. There is no inner debate as to "should I" or "should I not?" Similarly, the trend in the spiritual life of a Christian is in the direction of doing what's right because it is right.

Once we look to the Lord and head in the right direction, any shortfall will be covered by God's grace and Christ's perfection. Choosing to step in the right direction can appear intimidating, almost like climbing the Himalayas without conditioning, training or a backpack. Spiritual growth and survival Himalayas, requires absolute dependency on the power and guidance of the Creator.

That first step requires looking into the mirror, recognizing and turning away from what's ugly, and then putting it all on the line while praying, "God, have mercy on me, a sinner." [26]

"Born again" followers of God are invited to rest from physical labor one day each week to memorialize the creation miracle, and to commemorate freedom from physical bondage and celebrate delivery from sin's shackles.

"Trust in the Lord with all your heart and lean not on your own understanding; in all your ways acknowledge Him, and He will make your paths straight." [27]

Besides the inner renewal guaranteed to those who trust, the Creator also exemplifies compassion, forgiveness, justice and incomprehensible love. The Lord God Almighty sacrificed His own Son to salvage sinners. "For God so loved the world that He gave His one and only Son, that whoever believes in Him shall not perish but have eternal life." [28]

"In his great mercy he has given us new birth into a living hope through the resurrection of Jesus Christ from the dead, and into an inheritance that can never perish, spoil or fade---kept in heaven for you, who through faith are shielded by God's power until the coming of the salvation that is ready to be revealed in the last time." [29]

Those choosing to walk in God's straight paths will reflect the same light from the Spirit of God that banished the shroud of darkness enveloping earth at the beginning of creation week, *Three Days Before the Sun.*

"The power of one is exponential when linked to the power of The One." [30]

The epic story of Christ's birth, death, resurrection and promised return to restore perfection and to make all things new is inextricably linked to creation week. More than intellectual gymnastics debating theories, it comes down to the big picture perception of the truth about God.

Christianity is a unique religious faith in that it encompasses the truth about life's miraculous creation and its ultimate triumph over death.

Many who witnessed Christ's life, ministry, death, resurrection and ascension into heaven suffered martyrdom, secure in the promise Jesus would eventually return and award life eternal to all believers. Those steadfast believers who first broadcast the "good news" of hope didn't surrender their lives to perpetuate myth. They had seen with their own eyes--a crucified Christ conquer death.

They cherished the promise: "This same Jesus, who has been taken from you into heaven, will come back in the same way you have seen him go into heaven." [31]

*****

Christ is risen! He's alive!!!

Endnotes

Chapter I

1---Jonathan Wells, *Darwinism and Intelligent Design* (Washington, DC: Regnery Publishing,, Inc., 2006) 99, citing Stephen C. Meyer, "DNA and the Origin of Life: Information, Specification, and Explanation," 223-85. John Angus Campbell and Stephen C. Meyer (editors), *Darwinism, Design, and Public Education* (East Lansing, MI: Michigan State University Press, 2003).
2---Ernst Haeckel, *The Wonders of Life,* (London: Harper, 1905) 135 as cited by Stephen C. Meyer, *Signature in the Cell* (New York: Harper Collins Publishers, 2009) 44. Haeckel (1834-1919), is remembered for his corrupted embryo drawings, deliberately manipulated to support Darwin's chance hypothesis.
3---See Louis Pasteur, *Wikipedia*, the free encyclopedia.
4---Francis Crick, *Life Itself,* (New York: Simon and Schuster, 1981) 88.
5---Bill Gates, *The Road Ahead* (Boulder: Blue Penguin, 1996) 228; from "ID in PS Curricula," 1999.
6---Carl Werner, *Evolution: the Grand Experiment* (Green Forest, Arkansas: New Leaf Press, 2007) 193.
7---Michael Denton, *Evolution: A Theory in Crisis* (Chevy Chase, MD: Adler & Adler, Publisher, Inc., 1986) 338.
8---See endnote #1.
9---Mark Twain is credited with saying, *"If you don't read the newspaper you are uninformed, if you do read the newspaper you are misinformed."* Some media voices who are prone to question the authenticity of Intelligent Design or the wisdom of political candidates who advocate ID as a legitimate part of school curricula, may not have read the enlightened writings of Johnson, Behe, Dembski, Wells and Meyer.
10---Stephen C. Meyer, *Signature in the Cell* (New York: Harper One, 2009) *171,* citing Henry Quastler, *The Emergence of Biological Organization,* 16.
11---Michael Denton, *Evolution: A Theory in Crisis* (Bethesda, Md.: Adler & Adler, 1986) 331.
12---Sir Fred Hoyle, *The Universe: Past and Present Reflections* (University College, Cardiff, 1981) as referenced by Alan Hayward, *Creation and Evolution* (Minneapolis: Bethany House Publishers, 1995) 35.
13---Carl Werner, *Evolution: the Grand Experiment,* 195 citing Thaxton, C., Bradley, W., Olsen, R.., *The Mystery of Life's Origin: Reassessing Curent Theories* (Dallas: Lewis and StanleyPublishers, 1984) 163, 164.
14---Stephen C. Meyer, *Signature in the Cell, 201.* See also Jonathan Sarfati, "The Second Law of Thermodynamics: Answers to Critics," *Answers In Genesis,* (2002b), as cited by Burt Thompson and Brad Harrub, "15 Answers to John Rennie and *Scientific American's Nonsense*" (Montgomery, Alabama: Apologetics Press, Inc., 2002) 44, and A. Goffau, "Life With 482 Genes,"*Science.* 270 (5235), 445-6, 1995.
15---Lee M. Spetner, *Not by Chance* (Brooklyn, New York: The Judaica Press, Inc., 1997) 30.
16---J.C. Sanford, *Genetic Entropy (*Waterloo, New York: FMS Publications, 2008) 4.
17---Stephen C. Meyer, *Signature of the Cell,* 212.

Chapter II

1---Soren Lovtrup (Swedish biologist), *Darwinism: The Refutation of a Myth* (New York: Croom Helm, 1987), p. 422.
2---Charles Darwin, *On the Origin of Species by Means of Natural Selection, or the Preservation of Favored Races in the Struggle for Life,* (First Edition facsimile, 1859, Cambridge: Harvard University Press) 184.
3---*The Holy Bible*, King James Version. Genesis 1:21.
4---Charles Darwin, *Origin of Species,* 647.
5---__________, *Descent of Man,* Vol. I, 169.
6---__________, *Descent,* Vol. I, 168.
7---__________, *Descent,* Vol. 1, 216.
8---__________, *Descent,* Vol. I, 178.
8---__________, *Descent,* Vol. I, 201.
10---__________, *Descent,* Vol. II, 328.
11---__________, Descent, Vol. II, 327, 328.

13---David Quammen, *The Reluctant Mr. Darwin* (New York: Atlas Books, W.W. Norton & Company, 2006) 28.
14---Charles Darwin, *Origin,* 247.
15---__________, *Origin,* 453.
16--Bert Thompson, *The Scientific Case for Creation* (Montgomery, Alabama: Apologetic Press, Inc., 2002) 124.
17---Noble, et. al, *Parasitology,* sixth edition, "Evolution of Parasitism," Lea and Febiger, 1989, 516, as cited by Frank Sherwin, *Origins Issues,* "Natural Selection's Role in the Real World."
18---Colin Patterson, "Cladistics," *The Listener* (1982) 106:390.
19---Charles Darwin, Letter to J.D. Hooker [1 February] 1871, in Darwin, F., ed., *The Life and Letters of Charles Darwin,* [1898], (New York: Basic Books, Vol. II, 19590, reprint, 202-203.
20---Charles Darwin, *Descent,* Vol. II, 389.
21---__________, *Descent,* Vol. 1, 206.
22---__________, *Descent,* Vol. 1, 213.
23---Michael Denton, *Evolution: a Theory in Crisis,* (Bethesda, Maryland: Adler & Adler, 1986) 117, citing Darwin, C., (1858) in a letter to Asa Gray, 5 September, 1857, *Zoologist,* 16:1697-99, see 1699.
24---Charles Darwin, *Origin,* 649.
25---Charles Darwin to Asa Gray, cited by Adrian Desmond and James Moore, *Darwin,* (New York: W.W. Norton and Company, 1991) 456.
26---Desmond & Moore, *Darwin, 475.*
27---__________, *Darwin,* 477.
28---Sharon Begley, "How to Think Like a Scientist," *Newsweek,* July 9, 2007.

Chapter III

1---This thought is attributed to Mark Cahill, but the precise reference not known.
2---Martin Rees and Priyamvada Natarajan, "Invisible Universe," *Discover* (December, 2003) 18.
3---Charles Siebert, "Unintelligent Design," *Discover* (Vol. 27, No. 3, March, 2006) 31, 34.
4---Brian Greene, Professor of mathematics and physics, Columbia University, as quoted by Carl Warner, *Living Fossils, Evolution: the Grand Experiment,* (Green Forest, AR: New Leaf Press, 2008) 4.
5---Peter Coles, "Boomtime," *New Scientist,* March 3, 2007.
6---Dean L. Overman, 59, citing Hoyle and Wickramasinge, 148, 24, 150, 30, and 31 as quoted in Thaxton, Bradley, and Olsen, 196.
7---Harold J. Morowitz, *Energy Flow in Biology* (New York: Academic Press, 1968); cited by Coffin, 376.
8---John Keosian, In Haruhiko Nada, ed., *Origin of Life* (Tokyo: Center for Academic Publications, Japan Scientific Publications Press, 1978) 573, 574, quoted by Coffin, 377.
9---Walter L. Bradley and Charles B. Thaxton, "Information and the Origin of Life," in The Creation Hypothesis, ed. J.P. Moreland (Downers Grove, Illinois: Inter Varsity Press, 1994) 190 as quoted by Overman, 62.
10---Dean L. Overman, *A Case Against Accident and Self-Organization,* 58, 59.
11---Charles Darwin, *Origin,* 649.
12---__________, Letter to J.D. Hooker [1 February] 1871, in Darwin, F., ed., *The Life and Letters of Charles Darwin,* [1898], (New York: Basic Books, Vol. II, 19590) 202-203.
13---Dean L. Overman, *A Case Against Accident and Self-Organization* (New York: Rowman & Littlefield Publishers, Inc., 1997) 38. See Aleksander I. Oparin, *The Origin of Life* (New York: Dover Publications, Inc., 1938).
14---Charles B. Thaxton, Walter L. Bradley and Roger L. Olsen, *The Mystery of Life's Origin* (New York: Philosophical Library, 1984) 182, 183, 185, quoted by Bert Thompson, 80.30—Denton, 261.
15---Michael Denton, *Evolution, A Theory in Crisis,* 261.
16---Hubert P. Yockey, *Information Theory and Molecular Biology* (Cambridge: Cambridge University Press, 1992), 257, quoted by Overman, 61, 62.
17---Gunter Wachtershauser, Letter to Editor, *Science,* 25 October 2002, vol. 298.
18---Larry A. Witham, *Where Darwin Meets the Bible* (New York: Oxford University Press, 2002) 129.
19---Sir William Dawson, *The Story of Earth and Man* (New York: Harper and Brothers,

1887) 317, 322, 330, 339.
20---Paul Davies, *The Cosmic Blueprint: New Discoveries in Nature's Ability to Order the Universe* (New York: Simon & Schuster, 1988) 203 and *The Mind of God* (New York: Simon & Schuster, 1992) 232.
21---Michael J. Murray, *Reason for the Hope Within* (Grand Rapids, Michigan: Eerdmans, 1999) 61-62
22---__________, *Reason for the Hope Within,* 48.
23---Dean L. Overman, *A Case Against Accident and Self-Organization*, 40, 41.
24---Percival Davis, Dean H. Kenyon, and Charles B. Thaxton, Academic Editor, *Of Pandas and People* (Dallas: Haughton Publishing Company, 1993) 3, 4.
25---Ralph O. Muncaster, *Creation Versus Evolution* (Mission Viejo, Calif.: Strong Basis to Believe, 1997) 17.
26---Percival Davis, Dean H. Kenyon, and Charles B. Thaxton, Academic Editor, *Of Pandas and People,* 3.
27---Dean L. Overman, *A Case Against Accident and Self-Organization,* 42.
28---Michael Denton, *Evolution: A Theory in Crisis* (Bethesda, Md.: Adler & Adler,1986) 261, 262.
29---*The Holy Bible, New International Version* (Grand Rapids: Zondervan Bible Publishers, 1983), Genesis 1:2.
30---Rick Weiss, "Water Scarcity Prompts Scientists to Look Down," *Washington Post*, March 10, 2003, A-11.
31---Robert Jastrow, *Until the Sun Dies* (New York: W.W. Norton, 1977) 62, 63 as quoted by Bert Thompson, *The Scientific Case for Creation*, 76.
32---Chandra Wickramasinghe "Threats on Life of Controversial Astronomer," *New Scientist,* January 21, 1982, 140, as quoted by Overman, 60.
33---Thomas A. Edison, "One of the Many Thomas Edisons You Didn't Know About," *ARN Announce,* Access Research Network, February 14, 2011.
34---Wernher von Braun, as quoted by James Perloff, *Tornado in a Junkyard,* (Arlington, Massachusetts: Refuge Books, 1999) 253.

Chapter IV

1---Michael Denton, *Evolution: A Theory in Crisis*, (Bethesda, Md.: Adler & Adler, 1986) 96, 323.
2---Bert Thompson & Brad Harrub, "15 Answers to John Rennie and *Scientific American's* Nonsense," (Montgomery, Alabama: Apologetics Press, September, 2002) 31.
3---Charles Darwin, *Descent of Man,* Vol. II, 389.
4—Charles Darwin, Letter to J.D. Hooker [1 February] 1871, in Darwin, F., ed., *The Life and Letters of Charles Darwin,* [1898], (New York: Basic Books, Vol. II, 1959), 202-203.
5---Richard Hutton, "Evolution: The Series," *WashingtonPost.com, Live Online*, Wednesday September 28, 2001.
6---Charles Darwin, *Origin of Species,* 637.
7---Hubert P. Yockey, *Information Theory and Molecular Biology* (Cambridge: University Press, 1992), 257, quoted by Overman, 61, 62.
8---Michael Denton, *Evolution: A Theory in Crisis*, 250, 264.
9---Jonathan Sarfati, "The Second Law of Thermodynamics: Answers to Critics," answers In Genesis.org/docs/370.asp#crystals (2002b), cited by Burt Thompson & Brad Harrub, "15 Answers to John Rennie and *Scientific American's Nonsense*" (Montgomery, Alabama: Apologetics Press, Inc., 2002) 44.
10---Dean L. Overman, *A Case Against Accident and Self-Organization,* 63, 64.
11---Percival Davis, Dean H. Kenyon, and Charles B. Thaxton, Academic Editor, *Of Pandas and People* (Dallas: Haughton Publishing Company, 1993) 2.
12---Dean L. Overman, *A Case Against Accident and Self-Organization*, 40, 41.
13---Jonathan Sarfati, "The Second Law of Thermodynamics: Answers to Critics," answers In Genesis.org/docs/370.asp#crystals (2002b), cited by Burt Thompson & Brad Harrub, "15 Answers to John Rennie and *Scientific American's Nonsense*" (Montgomery, Alabama: Apologetics Press, Inc., 2002) 44.
14---Michael Denton, *Evolution: A Theory in Crisis.* 296, 323.
15---Percival Davis, Dean H. Kenyon, and Charles B. Thaxton, 2, 3.
16---Ashby L. Camp, The Myth of Natural Origins (Tempe, Arizona: Ktisisa Publishing, 1994) 31, 32.
17---Dean L. Overman, *A Case Against Accident and Self-Organization*, 45, 46.
18---Carl Werner, "Criticisms of the Stanley Miller Experiment." *Evolution: the Grand Experiment* (Green Forest, Arkansas: New Leaf Press, 2007) 207.

19---Thaxton, C., Bradley, W., Olsen, R., *The Mystery of Life's Origin: Reassessing Current Theories* (Dallas: Lewis and Stanley Publishers, 1984) 76, 77, as cited by Carl Werner, *Evolution: The Grand Experiment*, 249.
20---__________., *The Mystery of Life's Origin: Reassessing Current Theories* (Dallas: Lewis and Stanley Publishers, 1984) 81, as cited by Carl Werner, *Evolution: The Grand Experiment*, 207.
21---Carl Werner, *Evolution: The Grand Experiment*, 207, referencing Thaxton, C., Bradley, W., Olsen, R., *The Mystery of Life's Origin: Reassessing Current Theories* (Dallas: Lewis and Stanley Publishers, 1984) 91. The reference to *"3.9 billion years"* is within the context of conventional time calibrations.
22---Hubert P. Yockey, *Information Theory and Molecular Biology* (Cambridge: Cambridge University Press, 1992), 235, 236, 238, and 335 as quoted by Overman, 48.
23---"'Intelligent Design' smacks of Creation by Another Name," *USA Today*, Editorial Page, August 9, 2005.
24---Gearld A. Kerkut, *Implications of Evolution* (London: Pergamon, 1960) 6, 7.
25---__________, *Implications of Evolution*, 157.
26---Sir Fred Hoyle, "The Big Bang in Astronomy," *New Scientist* (November 19, 1981) 92:527, cited by Harrub and Thompson, "Creationists Fight Back."
27---Sir Francis Crick, *Life Itself: Its Origin and Nature* (New York, Simon & Schuster, 1981) 88, cited by Brad Harrub and Bert Thompson, "Creationists Fight Back," (Montgomery, Alabama: Apologetics Press, 2002) 2.
28---Ian T. Taylor, *In the Minds of Men* (Minneapolis: TFE Publishing, 1996), 161 and 182, quoting Pasteur as cited by Rene J. Dubos, *Louis Pasteur: Freelance of Science* (New York: Charles Scribner's Sons, 1976) 395.
29---Roddy M. Bullock, "Darwinists on Design: Jumping to Confusions," citing 1860 Darwin letter to Asa Gray, a designist, *The ID Report*, www.Discover.org,, 2-28-2009.
30---Michael Denton, *Evolution: A Theory in Crisis*, 264.
31---__________, *Evolution: A Theory in Crisis*. 342.
32---__________, *Evolution: A Theory in Crisis*, 290, 291.
33---Mark Twain, *Life on the Mississippi* (Boston: J.R. Osgood, 1883), 156 as quoted by Brad Harrub, *Reason and Revelation*, May, 2001, 21(5):38.

Chapter V

1---Henry Gee, *In Search of Deep Time* ((New York: The Free Press, 1999) 116, 117.
2—Charles Darwin, *Origin*, 219.
3---Charles Darwin, cited by F. Darwin, *The Life and Letters of Charles Darwin*, (1881), Vol. 3, 309.
4---Charles Darwin, *On the Origin of Species by Means of Natural Selection, or the Preservation of Favored Races in the Struggle for Life*, (First Edition facsimile, 1859, Cambridge: Harvard University Press) 184.
5---__________, *Descent, Vol. I*, 207
6---__________, *Descent,* Vol. II, 389.
7---__________, *Descent,* Vol. 1, 206
8---James Gibson, letter to Warren L. Johns (August 28, 1997); citing David Raup, *Zoologic Record* published in *Paleobiology* 2 (1976) 279-288.
9---A. G. Fisher, Grolier Multimedia Encyclopedia, 1998, fossil section.
10---See *Darwin's Dilemma*, an Illustra Media DVD production, 2009.
11---Kathy Sawyer, "New Light on a Mysterious Epoch," *The Washington Post* (February 5, 1998); Copyright 1998, The *Washington Post.*
12---Francisco J. Ayala and James W. Valentine, *Evolving, The Theory and Processes of Organic Evolution*, 1979, 266.
13---Peter Ward & Donald Brownlee, *Rare Earth*, Feb 2000, 150.
14---T.S. Kemp, *Fossils and Evolution*, (Oxford University, Oxford University Press, 1999) 253.
15—Henry Gee, *In Search of Deep Time* ((New York: The Free Press, 1999) 32.
16---Ian T. Taylor, "The Ultimate Hoax: Archaeopteryx Lithographica," *Proceedings of the Second International Conference on Creationism, Vol II* (Pittsburgh: Creation Science Fellowship, Inc., 1990) 279-291.
17---Alan Feduccia, T. Lingham-Soliar and J.R. Hinchliffe, *Journal of Morphology* (2005) 266(2): 125-166 as reported by David Coppedge, "Have We Been Sold a Bill of Goods About Dinosaurs and BirdEvolution?", *Creation Matters* (St. Joseph, Missouri: Creation Research Society, September/October, 2005) 6, 7.

18---*National Geographic*, November, 1999.
19---Darwin, *Origin,* 232.
20---Michael Denton, *Evolution: A Theory in Crises* (Bethesda, Maryland: Adler & Adler, 1986) 62, 358.
21---Gould, Stephen Jay (1980), "Is a New and General Theory of Evolution Emerging?," *Paleobiology*, 6[1]:119-130, Winter. This quote from Gould along with the following Patterson quote were cited by Jeff Miller, "Cro-Magnon Man: Nothing but a 'Modern Man,' " *Apologetics Press.Oef,* July 18, 2011.
22---Patterson, Colin (1979), Letter on April 10, 1979 to Luther Sunderland: reprinted in *Bible-Science Newsletter*, 19[8]:8, August, 1981.
23---Joseph Mastropaolo, "The Maximum-Power Stimulus Theory for Muscle," Creation Research Society Quarterly (St. Joseph, Missouri: Creation Research Society) Vol. 37, Number 4, March 2001, 213-219.
24-- Henry Gee, *In Search of Deep Time,* 116, 117.

Chapter IX

1---Richard Milton, *Shattering the Myths of Darwinism* (Rochester, Vt.: Park Street Press, 1997) 169, 170.
2--- Richard Lewontin, *The Triple Helix* (Cambridge, Massachusetts, HarvardUniversity Press, 2000) 91.
3---Lee Spetner, *Not by Chance: Shattering the Modern Theory of Evolution*, 97-103.
4---George Javor, "5,000 Years of Stasis," ("Genomic Science: 21st Century Threat to 19th Century Superstition," www.*CreationDigest.com*, Summer Edition, 2002).
5--See SanjaGupta, *CNN Health,* "Girl's mother just had 'feeling' something was wrong," *9-21-2010).*
6---Army's Edgewood Chemical Biological Center at Aberdeen Proving Grounds, Maryland, www.plosone.org/home.action, October 7, 2010.
7---Internet, www.beyondbooks.com. "Bacteria and Viruses."
8---Internet, http://en.wikipedia.org/wikl/Epigenetics.
9—Lee Spetner, *Not by Chance: Shattering the Modern Theory of Evolution*, 139, 141.
10---Kevin Anderson, "Radio Interview with Dr. Kevin Anderson," *Creation Matters*, No. 4 July/August 2004, 1.
11---Lee Spetner, *Not by Chance: Shattering the Modern Theory of Evolution*, 131, 141, 143.
12—Lee Spettner, *Not by Chance: Shattering the Modern Theory of Evolution*, 181, 198.
13---Michael Denton, *Evolution: A Theory in Crisis,* 334, 322.
14---Kevin Anderson, "Definition of Evolution," Anderson@nsric.ars.usda.gov, 9-4-2002, 1.
15---Marcia Barinaga, "Tracking Down Mutations That Can Stop the Heart ," *Science* 281 (July 3, 1998) 32.
16---Rick Weiss, "Defect Tied to Doubling of Risk for Colon Cancer," *The Washington Post*, August 26, 1997.
17---Tim Friend, "Gene Defect is Linked to Parkinson's," *USA Today* (June 27, 1997) and *USA Today*, January 17 (180, 2005.
18---Josie Glausiusz, "The Genes of 1996," *Discover* (January 1997) 36.
19--David A. Demick, "The Blind Gunman," *Impact* (El Cajon, Calif.: Institute for Creation Research, February, 1999) iv.
20--*The Star*, Ventura, California, June 24, 1997.
21---__________, June 24, 1997.
22---Reuters, "Genetic Error Causes Rapid-Aging Syndrome," *The Washington Post*, Thursday, April 17, 2003, A6.
23---Elizabeth Pennsi, "New Gene Found for Inherited Macular Degeneration," *Science* 281 (July 3, 1998) 31.
24---Daniel C. Weaver, "The River of Life," *Discover* (November 1997) 55.
25---Karen P. Steel and Steve D. M. Brown, "More Deafness Genes." *Science* 280 (May 29, 1998) 1403.
26---Rob Stein, "Sex May Rid Us of DNA Flaws," *The Washington Post* (February 1, 1999) A9.
27---See David Brown, "Scientists Discover 3 More Genes With Links to Alzheimer's Disease," *The Washington Post*, September 7, 2009, A3.
28---Mitch Lipka, *Consumer Ally,* "Meat Tainted With Deadly Virus is Being Sold to Consumers," September 28, 2010.
29---Liz Szabo, "Report: Just One Cigarette is Bad," *USA Today*, December 9, 2010, A-1.
30---Sean Pitman, *Detecting Design*.com, November 20, 2010.

31---Louis Bounoure *The Advocate*, 8 March 1984 , 17, quoted in *The Revised Quote Book*, 5. Bounoure has served as director of the Strasbourg Zoological Museum, and research director at the French National Center of Scientific Research.

32-Stephen J. Gould, Speech at Hobart College, February 14, 1980, cited by Luther Sunderland, *Darwin's Enigma* (El Cajon, California: Master Books, 1984), 106 (emphasis in original) cited by Bert Thompson and Brad Harrub, "*National Geographic* Shoots Itself in the Foot Again," (ApologeticsPress.Org online report, 2004) 36.

Chapter VI

1---*The Holy Bible,* New International Version, Genesis 1:20.

2---Charles Darwin, *Origin of Species*, 647, 453.

3---A. G. Fisher, *Grolier Multimedia Encyclopedia*, 1998, fossil section.

4---Peter Ward & Donald Brownlee, *Rare Earth*, Feb 2000, 150.

5---Kathy Sawyer, "New Light on a Mysterious Epoch," *The Washington Post* (February 5, 1998); Copyright 1998, The *Washington Post.*

6---Henry Gee, *In Search of Deep Time* (London: Comstock Publishing Associates, 1999) 116, 117

7---Henry Gee, *In Search of Deep Time,* 108, 110.

8---Charles Darwin, *Origin of Species,* 647.

9---Charles Q. Choi, "Cache in Chinese Mountain Reveals 20,000 Prehistoric Fossils," *LifeScienc.com,* December 22, 2010.

10---Charles Darwin, *Origin*, 647, 453.

11---Erik Stokstad, *Science,* 5 December 2003, 1645.

12---*Creation Matters* (Vol. 15, No. 6, November/December, 2010) 5, citing Kent State University (2010, November 9) as reported by www.physorg.com/news/2010-11.

13---Michael Denton, *Evolution: A Theory in Crisis.* (Bethesda, Md.: Adler & Adler, 1986) 298, 302.

14---Henry Gee, *In Search of Deep Time,* 133.

15---Steven A. Austin, ed., *Grand Canyon: Monument to Catastrophe* (Santee, California; Institute for Creation Research, 1994) 149.

16---Peter D. Ward, "Coils of Time," *Discover* (March 1998) 106.

17---Michael Denton, *Evolution: A Theory in Crisis.* (Bethesda, Md.: Adler & Adler,1986) 298, 302.

18--David Coppedge, "Speaking of Science," *Creation Matters,* January/February, 2010, 5.

19---Murray Wardrop, "Scientists Draw Squid Using its 150-million-year-old Fossilized Ink," *Telegraph.co.UK*, 19 August, 2009.

20---Charles Q. Choi, "Largest Fossil Spider Found in Volcanic Ash," Life Science.com as cited by *Yahoo! News*, April 21, 2011.

21---Charles Darwin, *Origin*, 219.

22---__________, *Origin*, 617, 618.

23---__________, *Origin*, 406.

24---__________, *Origin,* 232.

25---Henry Gee, *In Search of Deep Time*, 177, 155.

26---__________, *In Search of Deep Time*, 127.

27---Ernst Mayr, *The Growth of Biological Thought: Diversity, Evolution and Inheritance* (Cambridge, Massachusetts: The Belknap Press of Harvard University Press, 1982) 524.

28---Charles Darwin, *Origin*, 406.

29---R.L. Wysong, The Creation-Evolution Controversy. (Midland, Michigan: Inquiry Press, 1978) 348, 352-354.

30---Genesis 1:21, *The Holy Bible, New International Version.*

Chapter VII

1---Michael Denton, *Evolution: A Theory in Crisis.* 296, 323.

2---__________, *Evolution: A Theory in Crisis.* (Bethesda, Md.: Adler & Adler, 1986) 117, citing Darwin, C. (1858) in a letter to Asa Gray, 5 September, 1857, *Zoologist*, 16: 6297-99, see 6299.

3---__________, *Evolution: A Theory in Crisis.* 328, 329.

4---Charles Darwin letter to Asa Gray , cited by Adrian Desmond and James Moore, *Darwin,* (New York: W.W. Norton and Company, 1991) 456.

5---Desmond & Moore, *Darwin*, 475
6---Michael Denton, *Evolution: A Theory in Crisis*. 290, 291.
7---I. L. Cohen, *Darwin Was Wrong* (Greenvale, New York: New Research Publications, Inc., 1984) 38, 39, referencing H.G. Wells, Julian S. Huxley, and G.P. Wells, *The Science of Life* (New York: The Literary Guild) 41-43.
8---Michael Denton, *Evolution: A Theory in Crisis*. 328, 329.
9---Duane Arthur Schmidt, *And God Created Darwin* (Fairfax, Virginia: Allegiance Press, 2001) 24, 25.
10---I. L. Cohen, *Darwin Was Wrong* (Greenvale, New York: New Research Publications, Inc., 1984), 205.
11---Alan Hayward, *Creation and Evolution*. (Minneapolis: Bethany House Publishers, 1995) 35; referencing F. Hoyle, *The Universe: Past and Present Reflections* (University College, Cardiff, 1981).
12---I. L. Cohen, *Darwin Was Wrong* (Greenvale, New York: New Research Publications, Inc., 1984) 38, 39, referencing H.G. Wells, Julian S. Huxley, and G.P. Wells, *The Science of Life* (New York: The Literary Guild) 41-43.
13--Thomas Woodward, *Doubts About Darwin* (Grand Rapids: Baker Books, 2003) 44. Woodward referenced Sydney Fox in his review of *Mystery of Life's Origins in Quarterly Review of Biology*, June, 1985.
14---Harold Coffin, *Origin by Design* (Hagerstown, Maryland: Review and Herald Publishing Association, 1983) 379.
15---Joshua Lederberg, "A View of Genetics," *Science* 131 (3396) 1960: pp. 269-280 cited by Harold Coffin, *Origin by Design*, 377, 378.
16---Geoffrey Simmons, M.D., *What Darwin Didn't Know* (Eugene, Oregon: Harvest House Publishers, 2004) 43, 53.
17---Lee Spetner, *Not by Chance*. (Brooklyn, New York: The Judaica Press, Inc., 1997). 31.
18---Michael Denton, *Evolution: A Theory in Crisis*. 329.
19---__________, *Evolution: A Theory in Crisis*. 320, 321.
20---__________, *Evolution: A Theory in Crisis*. 342.
21---W. Wells, "Taking Life to Bits," *New Scientist*, 155(2095):30-33, 1997 as cited in "How Simple Can Life Be?", *AnswerwinGenesis.org*, 2009.
22---Walt Brown, *In the Beginning: Compelling Evidence for Creation and the Flood* (Phoenix, Ariz.: Center for Scientific Creation, 1996) 11, 12.
23---Carl Werner, *Evolution: the Grand Experiment*, 207.
24---Walt Brown, *In the Beginning: Compelling Evidence for Creation and the Flood* (Phoenix, Ariz.: Center for Scientific Creation, 1996) 11, 12.
25---W. Wells, "Taking Life to Bits," *New Scientist*, 155(2095):30-33, 1997 (see endnote #21)
26---Geoffrey Simmons, M.D., *What Darwin Didn't Know* (Eugene, Oregon: Harvest House Publishers, 2004) 43, 53.
27---Michael Denton, *Evolution: A Theory in Crisis*, 315.
28---Kirk R. Johnson and Richard E. Stucky, *Prehistoric Journey* (Boulder, Colorado: Roberts Rinehart Publishers, 1995) 16.
29---Dennis Wagner, "Top Ten Darwin and Design Science News Stories for 2010," ARN.org, /top10, December 21, 2010.
30---Albert Fleischmann, University of Erlangen Zoologist. See John Fred Meldau, ed., *Witnesses Against Evolution* (Denver: Christian Victory Publishing, 1968) 13.
31---Joyce Kilmer, *Trees and Other Poems* (New York, George H. Doran Company, 1914).

Chapter VIII

1---Carl Werner, *Evolution: the Grand Experiment* (Green Forest, Arkansas: New Leaf Press, 2007) 192, 193.
2--- Michael Denton, *Evolution: A Theory in Crisis*, 306.
3---__________, 345.
4---Michael Denton, *Evolution: A Theory in Crisis* (Bethesda, Md.: Adler & Adler, 1986) 331.
5--- I. L. Cohen, *Darwin Was Wrong*. (Greenvale, New York: New Research Publications, Inc., 1984) 40-42.
6---Carl Werner, *Evolution: the Grand Experiment* (Green Forest, Arkansas: New Leaf Press, 2007), 194.
7--- I. L. Cohen, *Darwin Was Wrong*, 209.
8---Sir Fred Hoyle, "The Big Bang in Astronomy," *New Scientist* (November 19,1981) 92:527, cited by Harrub and Thompson, "Creationists Fight Back."

9---This data is from an unidentified internet source. While believed to be true, its accuracy has not be verified independently.

10---Bill Gates, *The Road Ahead* (Boulder: Blue Penguin, 1996) 228; from "ID in PS Curricula," 1999.

11---Lee M. Spetner, . *Not by Chance*. Brooklyn, New York: The Judaica Press, Inc., 1997, 30.

12---George Javor, *CreationDigest.com*, Summer Edition, 2002.

13--- George Javor, "5,000 Years of Stasis," ("Genomic Science: 21st Century Threat to 19th Century Superstition," www.*CreationDigest.com*, Summer Edition, 2002).

14---I. L. Cohen, *Darwin Was Wrong*, 35, 54.

15---Michael Denton, *Evolution: A Theory in Crisis,* 334.

16---Joel Achenbach, "The Origin of Life Through Chemistry," *National Geographic*, March, 2006, 31.

17---Dean L. Overman, *A Case Against Accident and Self Organization* (New York: Rowman & Littlefield Publishers Inc., 1997) 59, citing Hoyle and Wickramasinge, 148, 24, 150, 30, and 31 as quoted in Charles B. Thaxton, Walter L. Bradley, and Roger Olsen, *The Mystery of Life's Origin: Reassuring Current Theories* (New York: Philosophical Library, 1984) 196.

18---Carl Werner, *Evolution: the Grand Experiment,* 195 citing Thaxton, C., Bradley, W., Olsen, R.., *The Mystery of Life's Origin: Reassessing Curent Theories* (Dallas: Lewis and StanleyPublishers, 1984) 163, 164.

19---________, 196-7.

20---Jonathan Sarfati, "The Second Law of Thermodynamics: Answers to Critics," *Answers In Genesis,* (2002b), as cited by Burt Thompson and Brad Harrub, "15 Answers to John Rennie and *Scientific American's Nonsense*" (Montgomery, Alabama: Apologetics Press, Inc., 2002) 44. See also, A. Goffau, "Life With 482 Genes,"*Science*. 270 (5235), 445-6, 1995.

21---Lee M. Spetner, *Not by Chance* (Brooklyn, New York: The Judaica Press, Inc., 1997) 30.

22---I. L. Cohen, *Darwin Was Wrong*., 205

23---Michael Denton, *Evolution: A Theory in Crisis*. 342.

24---Michael Denton, *Evolution: A Theory in Crisis*,, 338.

25---I. L. Cohen, *Darwin Was Wrong*., 40-42.

26---George Javor, "5,000 Years of Stasis," ("Genomic Science: 21st Century Threat to 19th Century Superstition," www.*CreationDigest.com*, Summer Edition, 2002).

27---Michael Denton, *Evolution: A Theory in Crisis*, 348, 357.

28---Richard Milton, *Shattering the Myths of Darwinism* (Rochester, Vt.: Park Street Press, 1997) 184; referencing Denton, *Evolution: A Theory in Crises.*

29--- I. L. Cohen, *Darwin Was Wrong*, 208.

30--- Michael Denton, *Evolution: A Theory in Crisis*, 324.

31---Antony Flew and Roy Varghese (2007), *There Is No God: How the World's Most Notorious Atheist Changed His Mind* (New York: HarperOne) 132.

Chapter IX

1--- Pierre-Paul Grassé, *The Evolution of Living Organisms,* (New York: Academic Press, 1977) 88, 103.

2---*The American Heritage Dictionary of the English Language,* Third Edition (New York, Houghton Mifflin Company, 1992) 754.

3--- Jobe Martin, *The Evolution of a Creationist* (Rockwall, Texas: Biblical Discipleship Publishers, 2002) 131, 132.

4---Ian Taylor, *In the Minds of Men*, 160, 161.

5--- Byron C. Nelson, *After Its Kind* (Minneapolis, Augsburg Publishing 1927) 98, 99.

6---Byron C. Nelson, *After Its Kind*, 101, referencing Alfred Russel Wallace, *Letters and Reminiscences*, 340.

7---Alfred Russel Wallace, "The Present Position of Darwinism," *Contemporary Review*, August, 1908.

8-- Jonathan Wells, *Icons of Evolution* (Washington, D.C.: Regnery Publishing, 2000) 180.

9--- Byron C. Nelson, *After Its Kind*, 101, quoting Wallace, *Letters and Reminiscences,* 95.

10--- Byron C. Nelson, *After Its Kind*, 101, citing *Theory of Evolution*, 163.

11---__________, 99,101 citing *Smithsonian Institute Report,* 1916, 343.

12---__________, 101, 102, citing *Science Progress*, January, 1925.

13---Albert Fleischmann, "The Doctrine of Organic Evolution in the Light of Modern Research," *Journal of the Transactions of the Victoria Institute* 65 (1933) 194-95, 205-6, 208-9.

14---A computerized comparison of the six editions can be checked out on Ben Fry's

website, “The Preservation of Favored Traces.”
15---Jonathan Wells, *Icons of Evolution*, 181.
16---Michael Denton, *Evolution: A Theory in Crisis* (Bethesda, Maryland: Adler & Adler, Publishers, inc., 1986) 75 citing Julian Huxley, *Evolution After Darwin* ed. Sol Tax, vol. 3 (Chicago: University of Chicago Press, 1960) 1-21, see 1.
17---See *A Scientific Dissent From Darwinism*, www.dissentfromdarwin.org.
18---Henry M. Morris, "What They Say," *Back to Genesis* (March 1999) a, b.
19---Lee Spetner, Not by Chance: Shattering the Modern Theory of Evolution (Brooklyn: Judaica Press, 1997) 160.
20--Bert Thompson, *The Scientific Case for Creation* (Montgomery, Alabama: Apologetic Press, Inc., 2002) 124.
21---Noble, et. al, *Parasitology,* sixth edition, “Evolution of Parasitism,” Lea and Febiger, 1989, 516, as cited by Frank Sherwin, *Origins Issues,* “Natural Selection’s Role in the Real World.”
22---Colin Patterson, “Cladistics,” *The Listener* (1982) 106:390.
23---Soren Lovtrup, *Darwinism: The Refutation of a Myth* (London: Croom Helm,1987) 352.

Chapter X

1---*The Holy Bible,* New International Version, Genesis 1:20.
2---Charles Darwin, *Origin of Species*, 647, 453.
3---A. G. Fisher, *Grolier Multimedia Encyclopedia*, 1998, fossil section.
4---Peter Ward & Donald Brownlee, *Rare Earth*, Feb 2000, 150.
5---Kathy Sawyer, "New Light on a Mysterious Epoch," *The Washington Post* (February 5, 1998); Copyright 1998, The *Washington Post.*
6---Henry Gee, *In Search of Deep Time* (London: Comstock Publishing Associates, 1999) 116, 117
7---Henry Gee, *In Search of Deep Time,* 108, 110.
8---Charles Darwin, *Origin of Species,* 647.
9---Charles Q. Choi, “Cache in Chinese Mountain Reveals 20,000 Prehistoric Fossils,” *LifeScienc.com,* December 22, 2010.
10---Charles Darwin, *Origin*, 647, 453.
11---Erik Stokstad, *Science,* 5 December 2003, 1645.
12---*Creation Matters* (Vol. 15, No. 6, November/December, 2010) 5, citing Kent State University (2010, November 9) as reported by www.physorg.com/news/2010-11.
13---Michael Denton, *Evolution: A Theory in Crisis.* (Bethesda, Md.: Adler & Adler, 1986) 298, 302.
14---Henry Gee, *In Search of Deep Time,* 133.
15---Steven A. Austin, ed., *Grand Canyon: Monument to Catastrophe* (Santee, California; Institute for Creation Research, 1994) 149.
16---Peter D. Ward, "Coils of Time," *Discover* (March 1998) 106.
17---Michael Denton, *Evolution: A Theory in Crisis.* (Bethesda, Md.: Adler & Adler,1986) 298, 302.
18--David Coppedge, “Speaking of Science,” *Creation Matters,* January/February, 2010, 5.
19---Murray Wardrop, “Scientists Draw Squid Using its 150-million-year-old Fossilized Ink,” *Telegraph.co.UK*, 19 August, 2009.
20---Charles Q. Choi, “Largest Fossil Spider Found in Volcanic Ash,” Life Science.com as cited by *Yahoo! News*, April 21, 2011.
21---Charles Darwin, *Origin*, 219.
22---__________, *Origin*, 617, 618.
23---__________, *Origin*, 406.
24---__________, *Origin,* 232.
25---Henry Gee, *In Search of Deep Time*, 177, 155.
26---__________, *In Search of Deep Time*, 127.
27---Ernst Mayr, *The Growth of Biological Thought: Diversity, Evolution and Inheritance* (Cambridge, Massachusetts: The Belknap Press of Harvard University Press, 1982) 524.
28---Charles Darwin, *Origin*, 406.
29---R.L. Wysong, The Creation-Evolution Controversy. (Midland, Michigan: Inquiry Press, 1978) 348, 352-354.
30---Genesis 1:21, *The Holy Bible, New International Version.*
31 ---Lyall Watson, “The Water People,” *Science Digest*, 90[5]:44, May, as cited by Eric

Lyons and Kyle Butt, *The Dinosaur Delusion*, (Montgomery, Alabama: Apologetics Press, Inc., 2008) 140.

Chapter XI

1---Joyce Kilmer, *Trees and Other Poems* (New York, George H. Doran Company, 1914).
2---Charles Darwin, *The Origin of Species,* (New York: Gramercy Books, 1859) 455. See Todd Charles Wood and Megan J. Murray, Understanding the Pattern of Life, (Nashvilled: Broadman and Holman Publishers, 2003) 15.
3---Tom Hennigan and Jerry Bergman, "The Origin of Trees," *Creation Research Society Quarterly*, Spring, 2011, Vol. 47, #4, 264. (see McLamb, E. and J.C. Hall, "The Quiet Evolution of Trees," 2010).
4---Henry Gee, *In Search of Deep Time* (London: Comstock Publishing Associates, 1999) 116, 117.
5---Tom Hennigan and Jerry Bergman, "The Origin of Trees," *Creation Research Society Quarterly*, Spring, 2011, Vol. 47, #4, 261.
6---Carl Werner, *Evolution: The Grand Experiment* (Green Forest, AR: New Leaf Press, 2007) 186.
7---Ulrich Kutschern and Karl J. Niklas, "Evolutionary Plant Physiology: Charles Darwin's Forgotten Synthesis," *Naturewissenschaften* (2009) 96: 1339-1354, 1340; Published online, September 18, 2009.
8---Darwin, C. (1881) in Darwin, F., *The Life and Letters of Charles Darwin,* (London: John Murray, 1888) vol. 3, 248; cited by Michael Denton, *Evolution: A* Theory in Crisis, 163.
9---R. Milner, *The Encyclopedia of Evolution: Humanity's Search for Its Origins* (New York: Facts on File Publishers, 1990) 14. See Charles Darwin, *Origin,* 647.
10---Eric Bapteste, a Pierre and Marie Curie University evolutionary biologist as quoted by "Science News," The Telegraph, Wednesday, July 13, 2001. Accessed 12 July 2011 through www.Telegraph.co.uk?science.
11---Tom Hennigan and Jerry Bergman, "The Origin of Trees," *Creation Research Society Quarterly*, Spring, 2011, Vol. 47, #4, 259, 268
12---C. Arnold, An Introduction to Paleobotany (New York: McGraw-Hill Publishing Company, 1947) 7, as cited by Carl Werner, *Evoluttion: the Great Experiment,* 186.
13---Henry Gee, *In Search of Deep Time,* 108, 110.
14---Tom Hennigan and Jerry Bergman, "The Origin of Trees," *Creation Research Society Quarterly*, Spring, 2011, Vol. 47, #4, 265. See also Johnson, H. *The International Book of Trees* (New York: Simon andSchuster, 1973) 24.
15---See *Creation ex Niblo*, December, 2000, 6.

Chapter XII

1---Michael J. Behe, *Darwin's Black Box.* (New York: The Free Press, 1996) 39.
2---__________, *Darwin's Black Box*, 15.
3---__________, *Darwin's Black Box*, 69-73.
4---__________, *Darwin's Black Box*, 31-36.
5---__________, *Darwin's Black Box,* 73.
6---__________, *Darwin's Black Box*, 79.
7---__________, *Darwin's Black Box*, 74-97.
8---__________, *Darwin's Black Box*, 86, 87.
9---__________, *Darwin's Black Box*, 90.
10---__________, *Darwin's Black Box*, 93, 94.
11---__________, *Darwin's Black Box*, 96, 97.
12---Michael Denton, "An Interview with Michael Denton," Access Research Network, Origins Research Archives, Vol. 15, Number 2, July 20, 1995.
13---__________, *Evolution: A Theory in Crisis.* (Bethesda, Md.: Adler & Adler, 1986) 117, citing Darwin, C. (1858) in a letter to Asa Gray, 5 September, 1857, *Zoologist*, 16: 6297-99, see 6299.
14--Donald R. Moeller, "Does a Smile Need 500 Million Years to Evolve?", *Creation Digest.com*, Spring Edition, 2002.
15--Albert Fleischmann, "The Doctrine of Organic Evolution in the Light of Modern Research," *Journal of the Transactions of the Victoria Institute 65* (1933) 194-95, 205-6, 208-9.
16---Frank Sherwin, "Un-Bee-lievable Vision," *Acts & Facts* (El Cajon, California: Institute

for Creation Research, Vol. 35, No. 2, February, 2006) 5.

17---Steve Austin, *Grand Canyon: Monument to Catastrophe* (El Cajon, California: Institute for Creation Research, 1994) 145 as cited by Frank Sherwin, "Un-Bee-lievable Vision," *Acts & Facts* (El Cajon, California: ICR, Vol. 35, No. 2, February, 2006) 5.

18---Sir William Dawson, *The Story of Earth and Man* (New York: Harper and Brothers, 1887) 317, 322, 330, 339.

19---Søren Løvtrup (Swedish biologist), Darwinism: The Refutation of a Myth (New York: Croom Helm, 1987) 422.

Chapter XIII

1---Wolfgang Smith, *Teilhardism and the New Religion* (Rockford, Ill.: TanBooks, 1988) 5,6, as quoted by James Perloff, *The Case Against Darwin* (Burlington, Massachusetts: Refuge Books, 2002) 71.

2---Gerald A. Kerkut, *Implications of Evolution*, 154.

3---See Carl Zimmer, "A New Step in Evolution," internet *Seed Design Series*, posted Jun 2, 2008.

4---Thomas Hayden, "A Theory Evolves," *U.S. News & World Report*, July 29, 2002, 42-44.

5---Lane P. Lester & Raymond G. Bohlin, *The Natural Limits to Biological Change* ,(Probe Books, 1989) 73.

6---Jonathan Wells, *Icons of Evolution* (Washington, D.C.: Regnery Publishing, Inc., 20000 182, 245. Dr. Wells holds Ph.D.s from both Yale University and the University of California at Berkeley.

7---Reported in *Science*, Vol. 295, 25, 11 Jan 2002.

8---David Brown, "Limits to Genetic Evolution," *The Washington Post,* July 7, 2003, A7.

9---Alan Feduccia, as quoted by Kathy A. Svitil, "Discover Dialogue," *Discover*, February, 2002, 16.

10---Elizabeth Culotta and Elizabeth Pennisi, "Evolution in Action," *Science*, 23 December 2005, Vol. 310, 1878-79.

11---See Maynard M. Metcalf in Adams, Leslie B., Jr., Editor/Publisher, *The Scopes Trial.* Birmingham, Alabama: The Legal Classics Library, 1984; a reprint of The World's Most Famous Trial, Third Edition (Cincinnati: National Book Company, 1925) 253.

12---Anthony Brown, Environment Editor, *The Times* (London, UK) 20 February 2003, reported by *info@creationresearch.net*, February 27, 2003.

13---Michael Denton, *Evolution: A Theory in Crisis*, 91.

14---Norman Macbeth, *Darwin Retried*, 36, citing Wilbur Hall, *Partner of Nature* (Appleton-Century, 1939).

15---Lane P. Lester & Raymond G. Bohlin, *The Natural Limits to Biological Change* (Dallas: Probe books, 1989) 95.

16---Adrian Higgins, "Why the Red Delicious No Longer Is," *The* Washington Post N ational Weekly Edition, August 15-21, 2005, 19.

17---Hans Ellegren, "Genomics: The Dog Has Its Day," *Nature*, 8 December 2005, 745, referenced in "Scientists Put Dogs on Genome Map," *USA Today* (Thursday, December 8, 2005) 8D.

18---Cheryl Adams Palmer, *Simple Love: A Short Essay About a Very Special Dog,* an unpublished article released on the internet, May 26, 2011.

19---Dennis Wagner, "2009 Annual Report: The Key Darwin and Design Science News Stories of the Year," *Access Research Network,* December 30, 2009.

20---Ernst Mayer,*The Growth of Biological Thought: Diversity, Evolution and Inheritance* (Cambridge, Massachusetts: The Belknap Press of Harvard University Press, 1982) 584.

Chapter XIV

1---Brian Thomas, "Fossilized Materials Must Be Young," *Acts & Facts,* June, 2009, 17.

2----Charles Darwin, *The Origin of Species,* 638.

3---Luigi Cavalli-Sforza, *Genes, Peoples, and Languages* (New York: North Point Press, 2000) 61 as cited by Eric Lyons and Kyle Butt, *The Dinosaur Delusion* (Montgomery, Alabama: Apologetics Press, Inc., 2008) 156.

4--- Creationwiki.org, "Carbon-14 Dating," 7 November, 2008.

5---Robert H. Brown, and C. L. Webster, "The Upper Limit of C-14 Age," *Origins*, Volume 15, 1988, 39.

6---Leonard Brand, 262, referencing P.A.L. Giem, *Scientific Theology,* (Riverside, California:

La Sierra University Press, 1997) 134-137.
7---John Baumgardner, with D. Russell Humphreys, Steven A. Austin,, and Andrew A. Snelling, "Measurable ^{14}C in Fossilized Organic Materials," *The Fifth International Conference on Creationism 2003*, 127; and reported in *Acts & Facts,* Vol. 32, No. 10, October 2003.
8---Ian T. Taylor, *In the Minds of Men.* Minneapolis, Minn.: TFE Publishing, 1991, 316.
9---Gunter Faure, *Principles of Isotope Geology* (Somerset, N.J.: John Wiley and Sons, Inc., 1986), 120, 121, 291; Copyright 1986, John Wiley & Sons, Inc. Reprinted by permission of John Wiley & Sons, Inc. as cited in Robert H. Brown letter to Warren L. Johns, October 22, 1995.
10---Richard Milton, *Shattering the Myths of Darwinism* (Rochester, Vermont:: Park Street Press, 1997) 53-55.
11---Robert H. Brown letter to Warren L. Johns, 26 November 2003.
12---Michael A. Cremo and Richard L. Thompson, *Forbidden Archeology* (Los Angeles: Bhaktivedanta Book Publishing, Inc., 1996) 694.
13---Andrew A. Snelling, "The Cause of Anomalous Potassium-Argon 'Ages' for Recent Andesite Flows at Mt. Ngauruhoe, New Zealand, and the Implications for Potassium-Argon Dating," *ICC Symposium Sessions* (Pittsburgh: Creation Science Fellowship, Inc., 1998) 510; *The Fifth International Conference onCreationism* (Pittsburgh: Creation Science Fellowship, Inc., 2003) 285-303.
14---Andrew A. Snelling, "Radioisotope Dating of Grand Canyon Rocks: Another Devastating Failure for Long-Age Geology," *Institute for Creation Research,* 2010.
15---Barry Yeoman, "Sweitzer's Dangerous Discovery," *Discover*, Vol. 27, No. 4, April, 2006, 37.
16---Brian Thomas citing V. Morel, "Dino DNA: the Hunt and the Hype. *Science.* 261 (5118): 160.
17---Unveil the "Holy Grail" of Paleontology in *Secrets of the Dinosaur Mummy.* Discovery Channel. Posted on discovery.com, accessed April 2, 2009, cited by Brian Thomas, "Fossilized Materials Must Be Young," *Acts & Facts,* June, 2009, 17.
18---Brian Thomas, citing D. Criswell, "How Soon Will Jurrasic Park Open? *Acts & Facts.* 35 (6).
19---"Dinosaur Mummy," *National Geographic News*, June 30, 2009.
20---Jason Palmer, "Dinosaur Mummy Yields Its Secrets," *BBC News,* June 30, 2009.
21---Charles Q. Choi, "Cache in Chinese Mountain Reveals 20,000 Prehistoric Fossils," *LifeScience.com,* December 22, 2010.
22---R.L. Wysong, *The Creation-Evolution Controversy.* (East Lansing, Michigan: Inquiry Press, 1976) 348.

Chapter XV

1---Denyse O'Leary, "One of the many Thomas Edisons you didn't know about," ARN Announce, Access Research Network, February 14, 2011.
2---*Holy Bible,* New International Version (Grand Rapids, Michigan, Zondervan Bible Publishers. 1978, Genesis 1:3.
3---__________, Genesis 1:2.
4---____________, I Corinthians 13:1
5---__________, Mathew 5: 14, 16.
6---__________, Genesis 1:14.
7---Eric Lyons and Kyle Butt, *The Dinosaur Delusion* (Montgomery, Alabama: Apologetic Press, Inc., 2008) 202.
8---*Holy Bible,* New International Version, Genesis 1:2.
9---__________, 1 Timothy 3:16.
10---__________, Deuteronomy 5:15
11---David Person, "150 Years Later, Civil War Redux," USA *Today,* 2-23-2011.
12---Stephen Budiansky, "Terror in the South," *Secrets of the Civil War* (U.S. News and World Report, 2008) 80.
13---*Holy Bible,* King James Version, Mark 2: 27.
14---See Brian Bull, Fritz Guy & Ervin Taylor, Editors, *Understanding Genesiss,* (Riverside, CA, Adventist Today, 2006).
15---*Holy Bible*, New International Version, Deuteronomy 4:20.

Chapter XVI

1---Michael Denton, *Evolution: A Theory in Crises* (Bethesda, Maryland: Adler & Adler, 1986) 358.
2---John Morris & Steven A. Austin, *Footprints in the Ash,* (Green Forest, Arkansas: Master Books, 2003) 67. See also, Steven A. Austin, "Excess Argon with Mineral Concentrations From the New Dacite Lava Dome at Mount St. Helens Volcano," *Creation Ex Nihilo Technical Journal*10 (1996), part 3; cited by *Acts and Facts,* Institute for Creation Research (May,1997) 26:5.
3---*The Holy Bible,* New International Version, Genesis 1:2.
4---Mark Swarts, "Scientists Confirm Age of the Oldest Meteorite Collision on Earth," *SpaceDaily.com*, August 23, 2002.
5---*The Holy Bible*, King James Version, Job 38:7.
6---F. Eisenhauer, et al., "A Geometric Determination of the Distance to the Galactic Center," *Astrophysical Journal,* (2003), 597.
7---Robert H. Brown letter to Warren L. Johns, 26 November 2003.
8---Frank Lewis Marsh, "On Creation with an Appearance of Age," *Creation Research Society Quarterly,* 1978, 14[4] 187, 188 as cited by Bert Thompson, *Creation Compromises* (Montgomery, Alabama: Apologetics Press, 2000) 270.
9---This summary is based upon the eyewitness account of Dr. Harold G. Coffin, paleontologist, who walked Surtsey, July, 1967.
10---*The Holy Bible,* New International Version, Hebrews 1:1, 2.
11---See Charles H. Hapgood, *Maps of the Ancient Sea Kings* (Kempton, Illinois: Adventures Unlimited Press, 1996) Preface, 193-197 & 244.
12---*Stephen B. Cox, "Atlantean-Egyptian Library," http:espah.tripod.com/europeanlibrary,* retrieved from internet, January 24, 2011.
13---*The Holy Bible,* New International Version, Genesis 11:3-7.
14---Eric Lyons, "Moses and the Art of Writing," ApologeticsPress.org/articles/779, March, 2010.
15---Samuel Wang and Ethel R. Nelson, *God and the Ancient Chinese* (Dunlap,Tennessee: Read Books, 1998) 31.
16---Geoffrey Barraclough, *TheTimes Atlas of World History* (London: Times Books Limited, 1978) 40, 52.
17---See *Smithsonian*, November, 2008, feature, Gobekli Tepe, Turkey.
18---See "Fossils in Antarctica," *British Antarctica Survey*, www.antarctica.ac.

Chapter XVII

1---*The Holy Bible*, New International Version, Romans 8:22.
2---John Anthony West, *The Serpent in the Sky,* (New York, NY: Harper and Row, 1984) as cited by David Hatcher Childress, *Technology of the Gods,* (Kempton, Illinois: Adventures Unlimited Press, 2000) 12.
3---Charles Darwin to Asa Gray, cited by Adrian Desmond and James Moore, *Darwin,* (New York: W.W. Norton and Company, 1991) *477.*
4---Yuri N. Ivanov, "Laws of Fertility, Role of Natural Selection, and Destructiveness of Mutations." *Creation Research Society Quarterly*, Vol. 17, December, 2000, 157.
5---Duane Arthur Schmidt, *And God Created Darwin* (Fairfax, Virginia: Allegiance Press, 2001) 131.
6---D. S. Allen and J. B. Delair, *Cataclysm* (Santa Fe, N. M.: Bear & Co., 1997) 107.
7---Elizabeth Kolvwer, "The Sixth Extinction?", *The New Yorker,* May 25, 2009, 57. See also "Mega Wombat Unearthed in Australia," *Yahoo News,* July 6, 2011.
8---*Creation*, 24(2):54, March-May, 2002.
9---See Ronald Pickering, *Nature Science Update*, 30 January 2002, online report Info@CreationResearch.net, February 14, 2002.
10--Dennis R. Peterson, *Unlocking the Mysteries of Creation* (El Dorado, California: Creation Resource Publications, 2002) 28.
11---Ariel A. Roth, *Origins.* (Hagerstown, Md.: Review and Herald Publishing Association, 1998) 182.
12---Doyle Rice, "Fossil Crowns 'King of the Rabbits,' " USA Today, March 31, 2011.
13---William Jacobs, "Goliath Squid by the Side of the Road," *Discover,* May, 2003, 16.

14---*Nature Science Update*, April 30, 2002.
15---*The Holy Bible*, New International Version, Genesis 6:4.
16---__________, Numbers 13:32-33.
17---For a detailed discussion see www.secondlaw.com/two.
18---Granville Sewell, "Evolution and the Second Law of Thermodynamics," www.isic.org/boards/ubb-get_topic-f-10-t-000038, January, 2004. Dr. Sewell serves in the Mathematics Department of Texas A&M University. Serious scholars deserve a look at Dr. Sewell's examination of evolution theory in the context of the second law of thermodynamics.
19---*Holy Bible*, New International Version, Genesis 3:4.
20---__________, Psalms 106:25, 26.

Chapter XVIII

1---Alfred Russell Wallace, *Darwinism* (London and New York, Macmillan and Co., 1890) 379, 380.
2---*The Holy Bible*, New International Version, Genesis 7:17-24.
3---*The Holy Bible,* King James Version, Genesis 1:2.
4---Mark Shwartz, "Scientists Confirm Age of the Oldest Meteorite Collision on Earth," *News Service*, (650) 723-9296; mshwarts@stanford.edu, August 20, 2002.
5---Gretel Schueller, "Australia's Ups and Downs," Earth (August 1998) 16.
6---Wilbur A. Nelson, *The Scopes Trial* (Birmingham, Ala.: The Legal Classics Library, 1984) a reprint of *The World's Most Famous Trial* (Cincinnati: National Book Co., 1925) 238-241.
7---Charles Darwin, *Origin of Species*, 648.
8---Richard Milton, *Shattering the Myths of Darwinism* (Rochester, Vermont: Park Street Press, 1997) 77, 78.
9---"Mystery of the Megaflood," Nova, 2005, pbs.org/previews/Nova_Megaflood.
10--Tom Vail, *The Grand Canyon: A Different View* (Green Forest, Arkansas: Master Books, 2003) 9.
11---Leonard Brand, *Faith, Reason, and Earth History.* Berrien Springs, Mich.: Andrews University Press, 1997) 254, 255.
12---Richard Milton, *Shattering the Myths of Darwinism* (Rochester, Vermont: Park Street
13---*The Holy Bible,* New International Version, Genesis 7:11, 19, and 20.
14---____________, Genesis 8:1.
15---____________, Genesis 6: 5-11
16---Lyall Watson, "The Water People," *Science Digest*, 90[5]:44, May, as cited by Eric Lyons and Kyle Butt, *The Dinosaur Delusion*, (Montgomery, Alabama: Apologetics Press, Inc., 2008) 140.
17---*The Holy Bible,* New International Version, Genesis 9:16.

Chapter XIX

1---John C. Whitcomb and Henry M. Morris, *The Genesis Flood* (Phillipsburg, New Jersey: Presbyterian and Reformed Publishing Company, 1995) 203.
2---Charles Q. Choi, "Cache in Chinese Mountain Reveals 20,000 Prehistoric Fossils," *LifeScience.com,* December 22, 2010.
3---Kathy Sawyer, "New Light on a Mysterious Epoch," *The Washington Post* (February 5, 1998).
4---Richard Milton, *Shattering the Myths of Darwinism*, 92.
5---__________, *Shattering the Myths of Darwinism*, 93.
6---Lyall Watson, "The Water People," *Science Digest*, 90[5]:44, May, as cited by Eric Lyons and Kyle Butt, *The Dinosaur Delusion*, (Montgomery, Alabama: Apologetics Press, Inc., 2008) 140.
7---Gretel Schueller, "Earth News: Death in the Dunes," *Earth* (June 1998) 11.
8---Luis Chiappe, "Dinosaur Embryos," *National Geographic* (December 1998) 38.
9---Ida Thompson, *National Audubon Society Field Guide to North American Fossils* (New York: Alfred A. Knopf, Inc., 1994) 765.
10--- Scott M. Huse, *The Collapse of Evolution* (Grand Rapids, Michigan: Baker, 1997) 96. A 1910 Geological Survey of Canada pictured a polystrate tree protruding vertically through multiple layers of sedimentary rock. (See photo, Ian T. Taylor, *In the Minds of Men* (Minneapolis: TFE Publishing, 1996) 114.
11---See Trevor Major, *Genesis & the Origin of Coal* (Montgomery, Alabama: Apologetics Press, 1996).

12--- John Paulien, *Armageddon at the Door* (Hagerstown, Maryland: Autumn House Publishing, 2008) 13.
13---See *The Washington Post Weekly Edition,* June 13-19, 2005, 10.
14---See Michael Guillen, *Five Equations that Changed the World*, 210.
15---David E. Shormann, "Novarupta and the Valley of 10,000 Smokes: Begging for a Biblical Interpretation," *Creation Research Society Quarterly,* Spring, 2010, 249.
16---Joel Achenbach, "When Yellowstone Explodes," *National Geographic,* August, 2009, 60, 61.
17---Larry Vardiman, "Are Hurricanes Getting More Destructive?", *Impact #390* (El Cajon, California: Institute for Creation Research, December, 2005) iv, referencing his previous research, L. Vardiman, 1996, *Sea-Floor Sediment and the Age of the Earth,* ICR Technical Monograph, Institute for Creation Research, EI Cajon, CA, 94 pp. and L. Vardiman., 2001, *Climates before and after the Genesis Flood: Numerical Models and Their Implications,* ICR Technical Monograph, Institute for Creation Research, EI Cajon, CA, 110 pp.
18---Michael J. Oard, *An Ice Age Caused by the Genesis Flood* (El Cajon, California: Institute for Creation Research, 1990) 33.
19---Michael J. Oard, *An Ice Age Caused by the Genesis Flood* (El Cajon, California: Institute for Creation Research, 1990) 34.
20---Tom Canby, "The Year Without a Summer," *Legacy* (Sandy Spring, Maryland: Sandy Spring Museum, 2002) Winter Edition.
21---Larry Vardiman, "Greenland Ice Cores," *CreationDigest.com.*, Winter, 2002; an updated version of "Impact Article #226" published by *The Institute forCreation Research* , April, 1992.
22---Margaret Olds, *Geologica* (New South Wales, Australia: Millennium House Pty Ltd., 2007) 402-3.
23---Dan Vergano, "Greenland Glacier Runoff Doubles Over Past Decade," *USA Today,* February 17, 2006, 24.
24---Rob Crilly, "Remote Somali Village Reels from Latest Hardship," *USA Today,* January 7, 2005, 5A.
25---See Chris Hawley, 'Researchers Explore Mysteries Surrounding 65-million-year-old Crater," *USA Today*, March 2, 2005, 9D.

Chapter XX

1---See Colin Patterson lecture, "Can You Tell Me Anything About Evolution," as transcribed by Wayne Frair and reported in "Bridge to Nowhere," *Creation Digest.com* website, Autumn 2004 Edition. Patterson, a lifelong evolutionist researcher, shared these and other doubts in an 1981 lecture before an audience of scientists at New York's American Museum of Natural History.
2-- Henry Gee, *In Search of Deep Time* (New York: The Free Press, 1999) 116, 117.
3---See *"Ida Fossil Discovery,"* The Darwinian Masillae, www.AgeoftheSage.org.
4---Randolph E. Schmid, "Before Lucy Came Ardi, New Earliest Hominid Found," *Yahoo News*, October 1, 2009.
5---Ann Gibbons, "Breakthrough of the Year, *Ardipithecus ramidus*," *Science*, December 18, 2009, Vol. 326, 1598-99.
6---Roddy M. Bullock, "Darwinists on Design: Jumping to Confusions," citing 1860 Darwin letter to Asa Gray, a designist, *The ID Report*, www.Discover.org, February 28, 2009.
7---Charles Darwin, *The Descent of Man,* Vol. II, 389.
8---__________, *Descent*, Vol. I, 203.
9---__________, *Descent*, Vol. II, 386.
10---__________, *Descent*, Vol. II, 389, 390.
11---__________, *Descent*, Vol. I, 207.
12---__________, Descent, Vol. II, 389.
13---__________, *Descent*, Vol. I, 206.
14---__________, Letter to Asa Gray, cited by Adrian Desmond and James Moore, *Darwin,* (New York: W.W. Norton and Company, 1991) 456.
15---__________, *Descent,* Vol. II, 328
16---__________, *Descent,* Vol. II, 327.
17---__________, *Descent*, Vol. I, 169.
18---__________, *Descent,* Vol. I, 168
19---__________, *Descent*, Vol. II, 216.
20---__________, *Descent,*, Vol. I, 201.

21---__________, *Descent*, Vol. I, 213.
22---__________, *Descent*, Vol. I, 173.
23---David Person, "150 years later; Civil War Redux," *USA Today,* February 23, 2011.
24---See Jamie Shreeve, "From Africa to Astoria by Way of Everywhere," *National Geographic*, September, 2009, 24.
25---*Creation Matters* (Vol. 15, No. 6, November/December, 2010) 7, citing Carnegie Museum of Natural History (2010, October 28, "Into Africa? Fossils Suggest Earliest Anthropoids Colonized Africa," *Science Daily*).
26---__________, citing R. Kaufman, "Oldest Modern Human Outside of Africa Found," *National Geographic News*, October 25, 2010.
27---"'Out of Africa' Theory Boost: Skull Dating Suggests Modern Humans Evolved in Africa," *Science Daily*, January 12, 2007.
28---See Richard M. Cornelius's definitive summary of the trial , "Scopes Trial: The Trial Gavel Heard Round the World," (*CreationDigest.com*, Winter, 2006) as excerpted from *Impact* (Dayton, Tennessee: Bryan College, 2000) v-xiii.
29---Fay Cooper-Cole, *The Scopes Trial* (Birmingham, Ala.: The Legal Classics Library, 1984) 237; a reprint of *The World's Most Famous Trial* [Cincinnati: National Book Co., 1925], 238-241.
30---George H. Hunter, *Civic Biology* (1914) 195, 196.
31---Lane P. Lester and Raymond G. Bohlin, *The Natural Limits to Biological Change* (Dallas: Probe Books, 1989) 54.
32---See *National Geographic*, "BODY: The Complete Human."
33---This list of the remarkable functions of the human organism is from the internet, author unknown. The veracity is presumed but is not authenticated.
34---See *The American Heritage Dictionary of the English Language,* some say thenumber exceeds 500,000).
35---*The Holy Bible,* King James Version, Genesis 8:4, 5. Clearly, angels were supernatural beings, created a little higher than *Homo sapiens.* Could secular speculative preoccupation with "aliens" inadvertently imply reference to angels of the Bible?
36---*The Holy Bible,* New International Version, Hebrews 13:2.
37---Jeff Miller, "Population Statistics and a Young Earth," *ApologeticsPress.org,* Jim Eastabrook, Editor, "Reason and Revelation," Volume 31, #5, May 9, 2011.
38---See "Mitochondrial Eve," Wikepedia.org referencing D.L Rohde, S. Olson and J.T. Chang, "Modeling the Recent Common Ancestry of All Living Humans," *Nature* (September, 2004) 431 and D.L. Rohde, "On the Common Ancestors of All Living Humans," submitted to *American Journal of Anthropology*, 2005.

Chapter XXI

1---Carolyn Leaf, *Who Switched Off My Brain?* (Southlake, TX: Inprov, 2009) 23.
2---See Dennis Normile, "Gene Expression Differs in Human and Chimp Brains," *Science*, 6 April 2001, 44, 45, presented at "Genes and Minds Initiative Workshop on Ape Genomics" in Tokyo, March 14-15, 2001.
3---See Duane T. Gish, *Evolution: The Fossils Still Say No!* (El Cajon, Calif.: Institute for Creation Research, 1995), 19, 20; Michael Denton, *Evolution: A Theory in Crisis* (Bethesda, Md.: Adler & Adler, 1986) 330, 331; and Harold Coffin with Robert H. Brown, *Origin by Design* (Hagerstown, Md.: Review and Herald Publishing Assn., 1983) 382.
4---Michael Denton, *Evolution: A Theory in Crisis* (Bethesda, Md.: Adler & Adler, 1986) 330.
5---C.P. Yu, "The Human Brain Testifies Against Evolution: Confessions of a Neurosurgeon," (Internet Website: www.hkam.org.hk/temp/counterevolution, as noted 1-10-2005).
6---Harold Coffin with Robert H. Brown, *Origin by Design* (Hagerstown, Md.: Review and Herald Publishing Association, 1983) 383.
7---Sharon Begley, "I can't Think," *Newsweek*, March 7, 2011, with apartial quote from Sheena Iyengar, *The Art of Choosing*.
8---*The Holy Bible*, King James Version, Proverbs 23:7.
9---Carolyn Leaf, *Who Switched Off My Brain?* (Southlake, TX: Inprov, 2009) 20.
10---__________, *Who Switched Off My Brain?* (Southlake, TX: Inprov, 2009) 29.
11---*The Holy Bible*, New International Version, Ephesians 4:31, 32.
12---_________, John 10:10.
13--Charles Darwin to T. Huxley, June 2, 1859, Desmond & Moore, *Darwin*, 475.
14---Adrian Desmond & James Moore, *Darwin*, (New York: W.W. Norton and Company, 1991) 477.

15---*The Holy Bible*, New International Version. Acts 17:24-26.

Chapter XXII

1---Wernher von Braun, as quoted by James Perloff, *Tornado in a Junkyard,* (Arlington, Massachusetts: Refuge Books, 1999) 253.
2---William B. Provine, Cornell University Professor of Biological Sciences summarizing "naturalistic evolution" in a 1998 Darwin Day keynote address, as per Sean Pittman's www.DetectingDesign.com.
3---See *National Geographic*, "Killer Stress," video.
4---Wernher von Braun, as quote by James Perloff, *Tornado in a Junkyard,* (Arlington, Massachusetts: Refuge Books, 1999) 253.
5---Ian T. Taylor, "The Idea of Progress," *The Fifth International Conference on Creationism,* (Pittsburgh: Creation Science Fellowship, Inc., 2003) 578.
6---*The Holy Bible,* New International Version,, Acts 7:60.
7---____________, Acts 8:3.
8---____________, Acts 9:1-18.
9---____________, 2 Thessalonians 2:3, 4.
10---An insightful observation cited by Faye Sandosky, original author unknown.
11---*The Holy Bible,* New International Version, Hebrews 12:1.

Chapter XXIII

1---Rick Warren, "Starbucks Stirs Things Up with a God Quote on Cups," *USA Today*, October 19, 2005, 8D.
2---See *The American Heritage Dictionary* (New York: Houghton Mifflin Co., 1992).
3---*The Holy Bible,* New International Version, II Timothy 3:5.
4---___________, Proverbs 14:12.
5---___________, Romans 1:20-25.
6---___________, Nehemiah 9:6.
7--*The Holy Bible*, New International Version, Revelation 12:7-9. These words of John the disciple and the revelator deserve credence. John walked at the side of Christ, an eyewitness to His life, death and resurrection. He devoted his own life to the founding of the Christian church and penned messages he learned from Christ's ministry.
8---___________, I Corinthians 4:9.
9---___________, Luke 12:6.
10---Wernher von Braun, as quoted by James Perloff, *Tornado in a Junkyard,* (Arlington, Massachusetts: Refuge Books, 1999) 253.
11---*The Holy Bible,* New International Version, Genesis 3:19.
12---___________, Genesis 3:17, 18.
13---*The Holy Bible,* New International Version, 2 Thessalonians 2:3-13.
14---___________, Revelation 1:7.

Chapter XXIV

1---Geroger Frederic Handel, *The Messiah* (1741-42).
2---*The Holy Bible,* New International Version, John 19:30.
3---___________, John 18:38.
4---___________, Matthew 24:24.
5---___________, I Corinthians 10:13.
6---___________, John 3:16
7---Augustinus, Aurelius (Augustine), *Confessions,* written between 397-398 AD, Albert C. Outlew's translation, revised by Mark Vessey (New York, Barnes and Noble, 2007). circa 397 AD, 183.
8--- *The Holy Bible,* New International Version, John 1:46.
9---___________, Matthew 4:19
10---___________,John 5:9.
11---*The Holy Bible,* King James Version, Mathew 5:3-10
12---*The Holy* Bible, New International Version, Mathew 5:14 & 15.
13---___________, John 15:13, 14, & 17.
14---_________, Luke 23d:34.

15---___________, I John 1:9.
16---This observation is credited to Mahatma Gandhi, a devotee of peace.
17---Carolyn Leaf, *Who Switched Off My Brain?* (Southlake, TX: Inprov, 2009) 22.
18---__________, *Who Switched Off My Brain?* (Southlake, TX: Inprov, 2009) 109.
19--- *The Holy Bible,* New International Version, Joshua 24:14, 15.
20---*The Holy* Bible, King James Version, Romans 6:23.
21--- *The Holy Bible,* New International Version, Psalms 145:8, 9.
22---___________, Psalms 51:10.
23---Sean Pitman, "Biology and Freewill," *www.DetectingDesign.com,* May 14, 2011.
24---*The Holy* Bible, King James Version, Philippians 4:7.
25---__________, *Matthew* 18:3.
26---__________, Luke 18:13.
27---__________, Proverbs 3:5, 6.
28---__________, John 3:16.
29---__________, I Peter 1:3-5.
30---Marilynn Peeke, "From the Pulpit," *Visitor*, October, 2009, 5.
31--- *The Holy Bible*, New International Version,, Acts 1:11

© Warren L. Johns

Sources

Sola Scriptura
One-thousand replicas of the most colorfully illuminated copies of the Gutenbert Bible were published in 1961.

The Bible's Old and New Testaments invest life with purpose and dimension, in sharp contrast to the superstitious imaginings suggesting first life organized itself from nothing, evolved without direction, only to end in the blackness of forever death.

More than a narrative of historical events, philosophy, poetry, and rules for better living, the 66-book compilation offers insight as to where we came from, what we are doing here and where we are going.

The Bible, the all-time best seller, speaks with inspired authority!

No ordinary literature, thoughts expressed came from *"men who spoke from God as they were carried along by the Holy Spirit."* (2 Peter 1:21).

An authoritative narrative of scores of miraculous events, the Bible preserves a panorama of history's unfolding miracle in action. Sixty-six books, authored by forty or so inspired humans from Moses to John the Revelator, writing between 1400 BC and 100 AD, provide reason for living and direction for every *Homo sapiens* who has ever lived.

Moses, an intellectual giant, who wrote from experience, insight and personal communication with God, is credited with having authored the Genesis account of life's beginning.

Blessed with access to the higher education reserved for Egyptian royalty, his faith in the Creator was absolute given his cultural exposure to the eyewitness and written accounts of creation week and the global deluge handed down by the likes of Adam, Seth, Noah, Shem, Abraham, Jacob and Joseph---devout worshipers of God.

John the Beloved, walked by Christ's side during His ministry, witnessed His crucifixion, and penned the Book of Revelation with a look to the future ---the last of the Bible's 66 books.

The "Big Picture" theme of Scripture begins with the miraculous origin of life on earth during the literal, seven-day creation week. Accounts of the life, death, and resurrection of Christ, God's only begotten son, tie the ends of historical events together in a living cavalcade of meaning and hope.

In the four centuries immediately following Christ's ministry, numerous letters were composed and circulated within the Christian fellowship---not all of which were anchored in the fundamentals of the Christian faith. Devoted scholars patiently sifted through the manuscripts and compiled the New Testament canon. The central message of the Messiah tied the Old Testament to the new in a single, cohesive entity.

Inspired insights of Moses, Daniel, David, Peter, Mathew, Mark, Luke, John and Paul, flow as complimentary components of a contiguous whole. Together, the Old and New Testaments provide purpose and meaning, for diverse cultures and every human life.

The Old Testament describes creation, the fall of man, the plan of redemption and the promise of a Messiah. The New Testament correlates with the Old by confirming the arrival of the Messiah, His message and ministry, His assured return, destruction of evil and an earth made new.

Moses, beneficiary of the finest education provided ancient Egyptian royalty, is credited with authoring Genesis, with its Creation account, and the four other books of the Pentateuch. An intellectual giant with a commanding presence, Moses lived without peer as lawgiver, courageous leader, and unselfish human dedicated to serving God, the Creator.

Peter fished in the Sea of Galilee until Christ called him to be a *"fisher of men."* Enthusiastic, visionary, loyal, courageous, and most of all a man of faith, Peter committed his life to teaching the truth about God; the life, death and resurrection of Christ, the life giver; and the miracle of creation.

On the road to Damascus, Saul of Tarsus, struck by the light of truth, did an about face. **Paul the Apostle,** turned about face from evil, devoting all his energies to presenting the Bible's "good news" of renewed life through Christ! A legend of living courage and faith, Paul put his own life on the line, preaching the truth about God and His creation miracle.

John the Revelator authored the gospel of John based on his personal eyewitness account of Christ's ministry, death and resurrection. Living until late in the first century A.D., John penned the Book of Revelation, alerting believers to the church's perilous road to end-time victory.

Augustine (Aurelius Augustinus, 354-430), Bishop of Hippo, converted to the Christian faith answering the prayers of his devout mother**,** played a pivotal role in merging the Old and New Testaments into the complete, *Holy Bible*. Thanks to influential church leaders like Augustine, aided by the translating skills of Jerome (384 A.D}, the complete Scriptural canon was formally ratified at the Synod of Hippo (393 A.D.)---nearly two centuries before Gregory, Bishop of Rome (590-604 A.D.) assumed the full religious and civil power of the Papacy.

John Wycliffe (1328-1384), fourteenth century Christian scholar, defied the religious establishment by translating the Bible canon into English in 1382. This courageous act of making the truth about God convenient to public accessibility so outraged empowered religionists that Wycliffe's bones were ordered exhumed from the grave and burned as a symbolic warning.

The **Waldenses** defied religious authoritarianism by going door-to-door sharing their understanding of Scripture. Branded as "heretics," more than 80 Waldensian martyrs were burned to death in Strasburg (1211 AD).

Johannes Gutenberg (1398-1468), printing press pioneer, published 180 copies of the Latin Vulgate version of the *Holy Bible* between the years 1450-1455. The Gutenberg Bible is honored as the first major book published using movable type.

Martin Luther (1483 -1546), German priest who championed salvation by faith alone, suffered excommunication from the Roman Catholic Churcb for challenging the pope's authority. Believing the Bible to be the primary source of religious truth, he translated the Latin scripture into the language of the people. Luther's credentials include credit as both an exceptional scholar as well as a courageous Christian leader pointing the way to the Reformation.

Five-hundred years later, the Bible is published, treasured and studied in most every world culture. No other book compares, either in content or in world-wide circulation just as no other faith-based religion worships an eternal, infinite God who created all life and demonstrated His exclusive power to conquor death as evident through the resurrection of Christ.

Achenbach, Joel, "When Yellowstone Explodes," *National Geographic,* August, 2009.

__________, "The Origin of Life Through Chemistry," *National Geographic*, March, 2006, 31.

Adams, Leslie B., Jr., Editor/Publisher, *The Scopes Trial* Birmingham, Alabama: The Legal Classics Library, 1984; a reprint of *The World's Most Famous Trial* (Third Edition) Cincinnati: National Book Company, 1925.

Age-of-the-Sage.org, "The Darwinius masillae."

Allan, D. S., and J. B. Delair. *Cataclysm.* Santa Fe, N.M.: Bea & Company, 1997.

Alter, Jonathan, "The President's 'Whiz Kids'", *Newsweek,* June 1, 2009.

American Heritage Dictionary of the English Language, The, Third Edition, New York: Houghton Mifflin Co., 1992.

Anderson, Kevin, "Radio Interview with Dr. Kevin Anderson," *Creation Matters,* No. 4 July/August 2004.

__________, "Definition of Evolution," Anderson@nsric.ars.usda.gov, 2002.

Appenzeller, Tim. "The Genes of 1996." *Discover,* January, 1997.

__________. "Test Tube Evolution Catches Time in a Bottle, "*Science* 284, June 25, 1999.

Army's Edgewood Chemical Biological Center at Aberdeen Proving Grounds, Maryland, www.plosone.org/home.action, **October 7, 2010.**

Arnold, C., *An Introduction to Paleobotany*, New York: McGraw-Hill Publishing Company, 1947.

Asp, Karen, "Rocks of Ages," *USA Today Open Air,* Spring, 2009.

Augustinus, Aurelius (Augustine), *Confessions,* written between 397-398 AD, Albert C.Outlew's translation, revised by Mark Vessey (New York, Barnes and Noble, 2007).

Austin, Steven A. "Excess Argon With Mineral Concentrates From the New Dacite Lava Dome at Mount St. Helens Volcano." *Creation Ex Nihilo Technical Journal,* vol. 10, part 3, 1996; as reported in "Acts and Facts," *Institute for Creation Research,* May, 1997.

__________, Editor. *Grand Canyon: Monument to Catastrophe,* .Santee, Calif.: Institute for Creation Research, 1994.

__________, "Interpreting Strata of Grand Canyon," in Austin, *Grand Canyon: Monument to Catastrophe*, Santee, California: Institute for Creation Research. 1994.

__________, "*Nautiloid Mass Kill and Burial Event, Redwell Limestone, Grand Canyon* Region," *Pittsburgh: International Conference on Creationism, 2003.*

Ayala, Francisco J. and James W. Valentine, *Evolving, The Theory and Processes of Organic Evolution, 1979.*

Barinaga, Marcia, "Tracking Down Mutations That Can Stop the Heart," *Science* 281, July 3, 1998.

Barraclough, Geoffrey, *The Times Atlas of World History* , London: Times Books Ltd, 1978.

Baumgardner, John with D. Russell Humphreys, Steven A. Austin,, and Andrew A.Snelling, "Measurable ^{14}C in Fossilized Organic Materials," *The Fifth International Conference on Creationism 2003,* 127; *Acts & Facts,* Vol. 32, No. 10, October 2003.

Begley, Sharon, "I can't Think," *Newsweek,* March 7, 2011, with apartial quote from Sheena Iyengar, *The Art of Choosing.*

Behe, Michael J. *Darwin's Black Box*, New York: The Free Press, 1996.

__________, *The Edge of Evolution,* New York, The Free Press, 2007.

Bird, W.R., *The Origin of Species Revisited,* Vol. I , Nashville: Regency, 1991.

Bounoure, Louis, *The Advocate,* 8 March 1984, quoted in *The Revised Quote Book.*

Bradley, Walter L. and Charles B. Thaxton, "Information and the Origin of Life," in *The Creation Hypothesis,* ed. J.P. Moreland, Downers Grove, Illinois: Inter Varsity Press, 1994.

Brand, Leonard. *Faith, Reason, and Earth History.* Berrien Springs, Mich.: Andrews University Press, 1997.

Brown, Anthony, *The Times,* London, UK, 20 February 2003, reported by *info@creationresearch.net,* February 27, 2003.

Brown, David, "Limits to Genetic Evolution," *The Washington Post,* July 7, 2003.

__________, "Scientists Discover 3 More Genes With Links to Alzheimer's Disease," *The Washington Post,* September 7, 2009, A3.

Brown, Robert H., "Amino Acid Dating," *Origins,* 1985, 12-8-25.

________, *Letter to Warren L. Johns,* 26 November, 2003.

Brown, Robert H, and C.L. Webster, "Interpretation of Radiocarbon and Amino Acid Data," Origins, 1991.

__________,"The Upper Limit of C-14 Dating," *Origins,* Volume 15, 1988.

Brown, Walt, *In the Beginning: Compelling Evidence for Creation and the Flood* Phoenix, Ariz.: Center for Scientific Creation, 1996.

Budd, Graham E. & Maximillian J. Telford, "Evolution: Along Came a Sea Spider," *Nature,* Vol. 437, October 20, 2005, as cited by Frank Sherwin, "Un-Bee-lievable Vision," *Acts & Facts* (El Cajon, California: ICR, Vol. 35, No. 2, February, 2006.

Budiansky, Stephen, "Terror in the South," *Secrets of the Civil War* (*U.S. News and World Report,* 2008) 80.

Buettner, Dan, *The Blue Zones,*(Washington, D.C., National Geographic, 2009.)
Bull, Brian, Fritz Guy & Ervin Taylor, Editors, *Understanding Genesiss,* (Riverside, CA, Adventist Today, 2006).
Bullock, Roddy M., "Darwinists on Design: Jumping to Confusions," citing 1860 Darwin letter to Asa Gray, a designist, *The ID Report,* www.Discover.org, February 28, 2009.
Camp, Ashby L., *The Myth of Natural Origins,* Tempe, Arizona: Ktisis Publishing, 1994 Edition.
Canby, Tom, "The Year Without a Summer," *Legacy,* Sandy Spring, Maryland: Sandy Spring Museum, 2002, Winter Edition.
Cavalli-Sforza, Luigi, *Genes, Peoples, and Languages,* New York: North Point Press, 2000.
Chiappe, Luis, "Dinosaur Embryos," *National Geographic,* December 1998.
Childress, David H., *Technology of the Gods,* (Kempton, Illinois: *Adventures* Unlimited Press,2000).
Choi, Charles Q., "Cache in Chinese Mountain Reveals 20,000 Prehistoric Fossils, *LiveScience.com,* December 22, 2010.
__________, "Largest Fossil Spider Found in Volcanic Ash," LifeScience.com as cited by *Yahoo! News,* April 21, 2011.
Coffin, Harold G., description of walk on beach, Surtsey, July, 1967.
Coffin, Harold, with Robert H. Brown, *Origin by Design.* Hagerstown, Md.: Review and Herald Publishing Association, 1983.
Cohen, I. L., *Darwin Was Wrong.* Greenvale, New York: New Research Publications, Inc., 1984.
Coles, Peter, "Boomtime," *New Scientist,* March 3, 2007.
Cooper-Cole, Fay, *The Scopes Trial*, Birmingham, Ala.: The Legal Classics Library, 1984, reprint of *The World's Most Famous Trial*, Cincinnati: National Book Co., 1925.
Coppedge, David, "Speaking of Science," *Creation Matters,* January/February, 2010.
Cornelius, Richard M., "Scopes Trial: The Trial Gavel Heard Round the World," as excerpted from *Impact* (Dayton, Tennessee: Bryan College, 2000, v-xiii.
Creation, 24(2):54, March-May, 2002.
***Creation Matters* (Vol. 15, No. 6, November/December, 2010).**
Creationwiki.org, "Carbon-14 Dating," 7 November, 2008.
Cremo, Michael A., and Richard L. Thompson. *Forbidden Archaeology.* Los Angeles: Bhaktivedanta Book Publishing, Inc., 1996.
Crick, Sir Francis, *Life Itself,* New York: Simon Schuster, 1981.
Crilly, Rob, "Remote Somali Village Reels from Latest Hardship," *USA Today,* January 7, 2005.
Culotta . Elizabeth and Elizabeth Pennisi, "Evolution in Action," Science, 23 December 2005, Vol. 310, 1878-79.
Darwin, Charles Robert, *On the Origin of Species by Means of Natural Selection, or the Preservation of Favored Races in the Struggle for Life,* First Edition facsimile, 1859, Cambridge: Harvard University Press.
__________, *The Origin of Species* (Sixth Edition), New York: Random House, Inc., 1993.
__________, *The Descent of Man, and Selection in Relation to* Sex, Vol. I & Vol. II Princeton, N.J., Princeton University Press, 1981.
__________, Letter to Asa Gray, cited by Adrian Desmond andJames Moore, *Darwin,* New York: W.W. Norton and Company, 1991.
__________, Letter to J.D. Hooker [1 February] 1871, in Darwin, F., ed., *The Life and Letters of Charles Darwin,* [1898], New York: Basic Books, Vol. II, 1959.
__________, (1881) in Darwin, F., *The Life and Letters of Charles Darwin,* London: John Murray, 1888, Vl. 3, cited by Michael Denton, *Evolution: A Theory in* Crisis.
Davies, Paul, *The Cosmic Blueprint: New Discoveries in Nature's Ability to Order the Universe,* New York: Simon & Schuster, 1988.
__________, *The Mind of God,* New York: Simon & Schuster, 1992.
Davis, Percival & Dean H. Kenyon, Charles B. Thaxton, Academic Editor, *Of Pandas and People,* Dallas: Haughton Publishing Company, 1993.
***Declaration of Independence*,** United States of America, July 4, 1776.
Demick, David A., "The Blind Gunman," *Impact* (El Cajon, Calif.: Institute for Creation Research, February, 1999) iv.
Desmond, Adrian & James Moore, *Darwin.* New York, Warner Books, Inc., 1991.
Denton, Michael, *Evolution: A Theory in Crisis.* Bethesda, Md.: Adler & Adler, 1986.
__________, "An Interview with Michael Denton," Access Research Network, Origins Research Archives, Vol. 15, Number 2, July 20, 1995.
Dawson, Sir William, *The Story of Earth and Man* (New York: Harper and Brothers, 1887).
Edison, Thomas A., "One of the Many Thomas Edisons You Didn't Know About," *ARN Announce,* Access Research Network, February 14, 2011.

Eisenhauer, F., et al., "A Geometric Determination of the Distance to the Galactic Center," *Astrophysical Journal,* (2003), 597.

Ellegren, Hans, "Genomics: The Dog Has Its Day," *Nature*, 8 December 2005, 745, referenced in "Scientists Put Dogs on Genome Map," *USA Today* ,Thursday, December 8, 2005.

Faure, Gunter, *Principles of Isotope Geology*. Somerset, N.J.: John Wiley and Sons, Inc., 1986.

Feduccia, Alan, as quoted by Kathy A. Svitil, "Discover Dialogue," *Discover*, February, 2002.

__________, **Feduccia, Alan, T. Lingham-Soliar and J.R. Hinchliffe,** *Journal of Morphology*, 2005, 266(2).

Fisher, A. G., *Grolier Multimedia Encyclopedia*, 1998.

Fleischman, Albert, "The Doctrine of Organic Evolution in the Light of Modern Research," *Journal of the Transactions of the Victoria Institute* 65, 1933.

Flew, Antony and Roy Varghese (2007), *There Is No God: How the World's Most Notorious Atheist Changed His Mind* (New York: HarperOne) 132.

Friend, Tim, "Gene Defect is Linked to Parkinson's," *USA Today*, June 27, 1997, and *USA Today*, January 17, 2005.

Gates, Bill, *The Road Ahead* (Boulder: Blue Penguin, 1996) 228; from "ID in PS Curricula," 1999.

Gee, Henry, *In Search of Deep Time: Beyond the Fossil Record to a New History of Life*, New York: The Free Press, 1999.

Gibbons, Ann, "Breakthrough of the Year, *Ardipithecus ramidus*," *Science*, December 18, 2009, Vol. 326.

Gibson, James, Letter to Warren L. Johns. August 28, 1997.

Gish, Duane T., *Evolution: The Fossils Still Say No!* El Cajon, Calif.: Institute for Creation Research, 1995.

__________, *Creation Scientists Answer Their Critics*. El Cajon, Calif.: Institute for Creation Research, 1993.

Gitt, Werner, *In the Beginning was Information*, Green Forest, Arkansas: Master Books, 2006.

__________, "Carbon-14 Content of Fossil Carbon," *Origins*, Number 51, 2001.

Glausiusz, Josie. "Fast Forward Aging." Discover, November, 1996.

__________. "The Genes of 1996." *Discover*, January 1997.

Gould, Stephen Jay, *The Panda's Thumb*, New York: W.W. Norton, 1980.

__________, "The Return of Hopeful Monsters, "*Natural History*, 86[4]:22-30, June-July, 1977.

__________, (1980), "Is a New and General Theory of Evolution Emerging?," *Paleobiology*, 6[1]:119-130, Winter.

__________, Speech at Hobart College, February 14, 1980, cited by Luther Sunderland, *Darwin's Enigma*, El Cajon, California: Master Books, 1984 cited by Bert Thompson and Brad Harrub, "*National Geographic* Shoots Itself in the Foot Again," Apologetics Press.Org online report, 2004.

Grassé, Pierre-Paul, *The Evolution of Living Organisms*,, New York:Academic Press, 1977.

Greene, Brian, as quoted by Carl Warner, *Living Fossils, Evolution: the Grand Experiment*, Green Forest, AR: New Leaf Press, 2008.

Guillen, Michael, *Five Equations that Changed the World*, New York: Hyperion, 1995.

Gupta, Sanjay, *CNN Health*, "Girl's mother just had "feeling Something Was Wrong," 9-21-2010).

Haeckel, Ernst, *The Wonderful Life*, J.McCabe translation, London: Harper, 1905.

Handel , George Frederic, *The Messiah* (1741-42).

Hapgood, Charles H. *Maps of the Ancient Sea Kings* (Kempton, Illinois: Adventures Unlimited Press, 1996).

Hayward, Alan, *Creation and Evolution*. (Minneapolis: Bethany House Publishers, 1995), referencing F. Hoyle, *The Universe: Past and Present Reflections* (University College, Cardiff, 1981).

Harrub, Brad and Bert Thompson, "Creationists Fight Back: A Review of *U.S. News & World Report's* Cover Story On Evolution," Montgomery, Alabama: Apologetics Press,2002.

Hawley, Chris, "Researchers Explore Mysteries Surrounding 65-million- year-old Crater," *USA Today*, March 2, 2005.

Hayden, Thomas, "A Theory Evolves," *U.S. News & World Report*, July 29, 2002.

Hennigan, Tom and Jerry Bergman, "The Origin of Trees," *Creation Research Society Quarterly*, Spring, 2011, Vol. 47, #4.

Higgins, Adrian, "Why the Red Delicious No Longer Is," *The Washington Post National Weekly Edition*, August 15-21, 2005.

Holy Bible, King James Version, New York: Oxford University Press; also New International Version, Grand Rapids: Zondervan Bible Publishers, 1978.

Hoyle, Sir Fred, "The Big Bang in Astronomy," *New Scientist*, November 19, 1981.

__________, *The Universe: Past and Present Reflections*. Cardiff: University College,1981.

Hoyle, Sir Fred and Chandra Wickramasinghe, *Evolution from Space*, London: J.M. Dent & Sons, 1981.

Hubbard, Ruth and Elijah Wald, *Exploding the Gene Myth*, Boston: Beacon Press1997, citing Francis Galton, *Inquiries Into Human Faculty*, London: Macmillan, 1883.

Humphreys, D. Russell Ph.D. with **John R. Baumgardner, Ph.D., Steven A. Austin, Ph.D.,** and **Andrew A. Snelling, Ph.D.,** "Helium Diffusion Rates Support Accelerated Nuclear Decay," *The Fifth International Conference on Creationism,* Robert L. Ivey, Jr., Editor, Pittsburgh: Creation Science Fellowship, 2003.
Hunter, Cornelius G., *Darwin's Proof,* Grand Rapids, Michigan: Brazos Press, 2003.
Huse, Scott M., *The Collapse of Evolution,* Grand Rapids, Michigan: Baker, 1997.
Huxley, Sir Julian, *Evolution After Darwin* ed. Sol Tax, vol. 3, Chicago: University of Chicago Press, 1960, the Centennial Celebration of the *Origin of Species.*
Holy Bible, King James Version, New York: Oxford University Press; also New International Version, Grand Rapids: Zondervan Bible Publishers, 1978.
Hoyle, Sir Fred, *The Universe: Past and Present Reflections* (University College, Cardiff, 1981)
Hoyle, Sir Fred and Chandra Wickramasinghe, *Evolution from Space,* London: J.M. Dent & Sons, 1981.
Hunter, George H., *Civic Biology* , 1914.
Hutton, Richard, "Evolution: The Series," *Washington Post.com, Live Online,* Wednesday, September 28, 2001.
Internet, www.beyondbooks.com. "Bacteria and Viruses."
Internet, http://en.wikipedia.org/wikl/Epigenetics.
Ivanov, Yuri N., "Laws of Fertility, Role of Natural Selection, and Destructiveness of Mutations." *Creation Research Society Quarterly,* Vol. 17, December, 2000.
Jacobs, William, *"Goliath Squid by the Side of the Road,"* Discover, *May, 2003.*
Jastrow, Robert, *Until the Sun Dies,* New York, W.W. Norton, 1977, as quoted by Bert Thompson, *The Scientific Case for Creation.*
Javor, George, "5,000 Years of Stasis," "Genomic Science: 21st Century Threat to 19t Century Superstition," www.*CreationDigest..com,* Summer Edition, 2002.
Johns, Warren LeRoi. *Ride to Glory,* Brookeville, Maryland: General Title, Inc., 1999.
__________, *Beyond Forever,* Smithville, Tennessee: *Creation Digest, 2007.*
__________, *Genesis File,* Smithville, Tennessee, www.GenesisFile.clom, 2010.
__________, *Chasing Infinity,* Smithville, Tennessee, www.GenesisFile.clom, 2011.
__________, *Time Zero,* Smithville, Tennessee, www.GenesisFile.clom,
Johnson, Kirk R. and **Richard K. Stucky** *Prehistoric Journey,* Boulder, Colorado: Roberts Rinehart Publishers, 1995.
Johnson, Philip E., *Darwin on Trial,* Washington, D.C., Regnery Gateway, 1991.
Kemp, T.S.,, *Fossils and Evolution, Oxford University, Oxford University Press,* 1999.
Keosian, John, In Haruhiko Nada, ed., *Origin of Life,* Tokyo: Center for Academic Publications, Japan Scientific Publications Press, 1978.
Kerkut, G.A., *Implications of Evolution,* New York: Pergamon Press, 1965.
Kilmer, Joyce, *Trees and Other Poems* (New York, George H. Doran Company, 1914).
Kime, Wesley, internationally recognized portrait artist, writes with the academic credentials of a Physician, finished at the top of his 1953 School of Medicine class, 2008 statement to Editor.
King, Colbert I., "A Dangerous Kind of Hate," *The Washington Post,* September12, 3009.
Kilbert, Elizabeth, "The Sixth Extinction?", *The New Yorker,* May 25, 2009.
Kutschern, Ulrich and Karl J. Niklas, "Evolutionary Plant Physiology: Charles Darwin's Forgotten Synthesis," *Naturewissenschaften* (2009)96: 1339-1354, published online, September 18, 2009.
Leaf, Caroline, *Who Switched Off My Brain?* (Southlake, TX, InProcv, 2009).
Lederberg Joshua, "A View of Genetics," *Science* 131 (3396) 1960: pp. 269-280 cited by Harold Coffin, *Origin by Design.*
Leno, Jay, "Jay Leno and 3D Printers," Internet email, July 10, 2011.
Lester, Lane P., and **Raymond G. Bohlin.** *The Natural Limits to Biological Change* Dallas, Texas: Probe Books, 1989.
Lewontin, Richard , *The Triple Helix,* Cambridge, MS, Harvard University Press, 2000.
Lipka, Mitch,*Consumer Ally,* "Meat Tainted With Deadly Virus is Being Sold to Consumers," September 28, 2010
Løvtrup, Søren. *The Refutation of a Myth,* New York: Croom Helm, 1987.
Lubenow, Marvin L., *Bones of Contention.* Grand Rapids, Michigan: Baker Books,1992, 2004.
Lyons, Eric and Kyle Butt, *The Dinosaur Delusion,* Montgomery, Alabama: Apologetic Press, 2008.
Macbeth, Norman. *Darwin Retried: An Appeal to Reason.* Boston: The Harvard Common Press, 1978.
Mackay, John, "Evidence News Update No. 4, *Creation Research,* April 2, 2003.
Major, Trevor, *Problems in Radiometric Dating,* Research Article Series, Montgomery, Alabama: Apologetics Press, Inc., undated publication.
Marsh, Frank Lewis. "On Creation with an Appearance of Age," *Creation Research Society Quarterly,* 1978, 14[4].

Martin, Jobe, *The Evolution of a Creationist* , Rockwall, Texas: Biblical Discipleship Publishers, 1994.
Mastropaolo, Joseph, "The Maximum-Power Stimulus Theory for Muscle, "Creation Research Society Quarterly (St. Joseph, Missouri: Creation, Research Society) Vol. 37, No. 4, March 2001.
Mayr, Ernst. *Systematics and the Origin of Species.* New York: Columbia University Press, 1942; Dover Publications paperback, 1964.
__________, *The Growth of Biological Thought: Diversity, Evolution and Inheritance,* Cambridge, Massachusetts: The Belknap Press of Harvard *University* Press, 1982.
McElheny, Victor K. *Watson and DNA,* Cambridge, Massachusetts: Perseus Publishing, 2003.
Metcalf, Maynard M., in Adams, Leslie B., Jr., Editor/Publisher, The Scopes Trial Birmingham, Alabama: The Legal Classics Library, 1984; a reprint of The World's Most Famous Trial, Third Edition, Cincinnati: National Book Company, 1925.
Meyer, Stephjen C., *Signature in the Cell,* New York: Harper Collins, 2009.
__________,"DNA and the Origin of Life: Informationk, Specification, and Explanation," 223-85. John Angus Campbell and Stephen C. Meyer (editors), *Darwinism, Design, and Public Education* (East Lansing, MI: Michigan State University Press, 2003).
Miller, Jeff, "Population Statistics and a Young Earth," *ApologeticsPress.org,* Jim Eastabrook, Editor, "Reason and Revelation," Volume 31, #5, May 9, 2011.
Milner, R., *The Encyclopedia of Evolution: Humanity's Search for Its Origins,* New York: Facts on File Publishers, 1990.
Milton, Richard. *Shattering the Myths of Darwinism.* Rochester, Vermont.: Park Street Press, 1997.
Moeller, Donald R., "Does a Smile Need 500 Million Years to Evolve?", *CreationDigest.com,* Spring Edition, 2002.
Morris, Henry M., "The Microwave of Evolution," *Back to Genesis,* August, 2001, a.
__________. "What They Say." *Back to Genesis.* N. Santee, CA: Institute for Creation Research, 1999.
Morris, John & Steven Austin, *Footprints in the Ash,* Green Forest, Arkansas: Master Books, 2003.
Morowitz, Harold J., *Energy Flow in Biology,* New York: Academic Press, 1968.
Murray, Michael J., *Reason for the Hope Within,*Grand Rapids, Michigan: Eerdmans, 1999.
National Geographic, November, 1999; and promo, "BODY: The Complete Human."
National Geographic News, "Dinosaur Mummy," June 30, 2009.
Nature Science Update, April 30, 2002.
Nelson, Byron C. *After Its Kind.* Minneapolis: Augsburg Publishing House, 1927.
__________, referencing Alfred Russell Wallace, *Letters and Reminiscences.*
Nelson, Wilbur A., *The Scopes Trial* , Birmingham, Ala.: The Legal Classics Library, 1984.
Normile, Dennis, "Gene Expression Differs in Human and Chimp Brains," *Science,* 6 April 2001, presented at "Genes and Minds Initiative Workshop on Ape Genomics" in Tokyo, March 14-15, 2001.
Nova, "Mystery of the Megaflood," pbs.org/previews/Nova_Megaflood., Nova, 2005,
Oard, Michael J., "The Origin of the Grand Canyon, Part II, Fatal Problems with the Dam-Breach Hypothesis," *Creation Research Society Quarterly,* Spring, 2010, 290.
__________, *An Ice Age Caused by the Genesis Flood* El Cajon, California: Institute for Creation Research, 1990.
Olasky, Marvin and John Perry, *Monkey Business,* Nashville: Broadman & Holman, Publishers, 2005.
Olds, Margaret, *Geologica,* New South Wales, Australia: Millennium House Pty Ltd., 2007.
Oliver & Boyd, *Contemporary Botanical Thought,* 1971.
Overman, Dean L., *A Case Against Accident and Self-Organization.* New York: Rowman & Littlefield Publisher, Inc., 1997.
Palmer, Cheryl Adams, *Simple Love: A Short Essay About a Very Special Dog,* an unpublished article released on the internet, May 26, 2011.
Palmer, Jason, "Dinosaur Mummy Yields Its Secrets," *BBC News,* June 30, 2009.
Patterson, Colin, "Can You Tell be Anything About Evolution," lecture as transcribed by Dr. Wayne Frair and reported in "Bridge to Nowhere?", *CreationDigest.com,* Autumn 2004 Edition.
__________, Letter on April 10, 1979 to Luther Sunderland: reprinted in *Bible-Science Newsletter,* 19[8]:8, August, 1981.
Paulien, John, *Armageddon at the Door* , *Hagerstown, Maryland: Autumn House Publishing, 2008.*
Peeke, Marilynn, "From the Pulpit," *Visitor,* October, 2009.
Person, David, "150 Years Later, Civil War Redux," USA *Today,* 2-23-2011.
Peterson, Dennis R. *Unlocking the Mysteries of Creation* (El Dorado, California: Creation Resource Publications, 2002) 28.
Pennisi, Elizabeth. "Genome Data Shakes Tree of Life." *Science* 280, May 1, 1998.
__________, "New Gene Found for Inherited Macular Degeneration." *Science* 281, July 3, 1998.
Perloff, James. *Tornado in a Junkyard,* Arlington, Mass.: Refuge Books, 1999.
Pickering Ronald, *Nature Science Update,* 30 January 2002, online report **Info@CreationResearch.net**, February 14, 2002.

Pitman, Sean, *Detecting* Design.com, November 20, 2010l.
__________, "Biology and Freewill," *www.DetectingDesign.com,* May 14, 2011.

Provine, William B., 1998 Darwin Day keynote address, as reference by Sean Pittman in his website, www.DetectingDesign.com.

Quammen, David, *The Reluctant Mr. Darwin,* New York: Atlas Books, W.W. Norton & Company, 2006.

Quastler, Henry, *The Emergence of Biological Organization,* New Haven, CT: Yale University Press, 1964.

Rappuoli, Rino, Henry L. Miller, and **Stanley Falkow,** "The Intangible Value of Vaccination," *Science* Vol. 297, 9 August 2002.

Rees, Martin and **Priyamvada Natarajan,** "Invisible Universe," *Discover,* December, 2003.

Rohde, D.L and S. Olson and J.T. Chang, "Modeling the Recent Common Ancestry of All Living Humans," *Nature* (September, 2004) 431 and D.L. Rohde, "On the Common Ancestors of All Living Humans," submitted to *American Journal of Anthropology,* 2005.

Rice, Doyle, "Fossil Crowns 'King of the Rabbits,' " USA Today, March 31, 2011.

Roth, Ariel A. *Origins.* Hagerstown, Md.: Review and Herald Publishing Assn., 1998.

Ruse, Michael, *The Evolution-Creation Struggle* , Cambridge: Harvard University Press, 2005.

Ryan, William and **Walter Pitman,** *Noah's Flood.* New York: Simon & Schuster, 1998.

Sanford, J.C., *Genetic Entropy and the Mystery of the Genome,* Waterloo, New York, FMS Publications, 2008.

Sarfati, Jonathan, *Refuting Evolution.* Green Forest, Arkansas: Master Books, 1999.
__________, "The Second Law of Thermodynamics: Answers to Critics," answersingenesis.org/docs/370.asp#crystals (2002b).

Sawyer, Kathy. "New Light on a Mysterious Epoch," *The Washington Post,* February 5, 1998.

Schmidt, Duane Arthur, *And God Created Darwin* (Fairfax, Virginia: Allegiance Press, 2001).

Scmid, Randolph E., "Before Lucy Came Ardi," *Yahoo News,* October 1, 2009.

***Science*,** Vol. 295, 25, 11 Jan 2002.

***Science Daily*,** January 12, 2007 "'Out of Africa' Theory Boost: Skull Dating Suggests Modern Humans Evolved in Africa."

Schueller, Gretel. "Australia's Ups and Downs." *Earth,* August, 1998.
__________, "Earth News: Death in the Dunes," *Earth,* June, 1998.

Siebert, Charles, "Unintelligent Design," *Discover,* March, 2006.

Sewell, Granville, "Evolution and the Second Law of Thermodynamics," www.isic.org, January, 2004.

Shelton, J. S., *Geology Illustrated* , San Francisco and London:W. H. Freeman and Co.

Sherwin, Frank, "Un-Bee-lievable Vision," *Acts & Facts,* El Cajon, California: Institute for Creation Research, Vol. 35, No. 2, February, 2006.
__________, *Origins Issues,* "Natural Selection's Role in the Real World," citing Noble, et. al, *Parasitology,* sixth edition, "Evolution of Parasitism," Lea and Febiger, 1989.

Shormann, David E., "Novarupta and the Valley of 10,000 Smokes: Begging for a Biblical Interpretation," *Creation Research Society Quarterly,* Spring, 2010, 249.

Shreeve, Jamie, "From Africa to Astoria by Way of Everywhere," *National Geographic,* September, 2009, 24.

Shwartz, Mark, "Scientists Confirm Age of the Oldest Meteorite Collision on Earth," *News Service,* (650) 723-9296; mshwarts@stanford.edu, August 20, 2002.

Simmons, Geoffrey, M.D., *What Darwin Didn't Know,* Eugene, Oregon: Harvest HousePub., 2004.

Smith, Shelton, "What We Believe," *www.SwordoftheLord.com.*

Smith, Wolfgang, *Teilhardism and the New Religion.* Rockford, Ill.: Tan Books, 1988.

Spetner, Lee M. *Not by Chance.* Brooklyn, New York: The Judaica Press, Inc., 1997.

Snelling, Andrew A., "The Cause of Anomalous Potassium-Argon 'Ages' for Recent Andesite Flows at Mt. Ngauruhoe, New Zealand, and the Implications for Potassium-Argon Dating." *ICC Technical Symposium Sessions,* Robert E.Walsh, Editor.Pittsburgh: Creation Science Fellowship, Inc. 1998.
__________, "Radioisotope Dating of Grand Canyon Rocks: Another Devastating Failure for Long-Age Geology," *Institute for Creation Research,* 2010.

***Star, The*,** Ventura, California, June 24, 1997.

Steel , Karen P. and Steve D. M. Brown, "More Deafness Genes." *Science* 280, May 29, 1998.

Stein, Rob, "Sex May Rid Us of DNA Flaws," *The Washington Post* ,February 1, 1999, A9.

Stokstad, Erik, "Gutsy Fossil Record for Staying the Course," *Science,* Vol. 302, 5 December 2003.

Sunderland, Luther, *Darwin's Enigma: Fossils and Other Problems,* San Diego: Master Books, 1988.

Swarts, Mark,"Scientists Confirm Age of the Oldest Meteorite Collision on Earth, "*SpaceDaily.com,* August 23, 2002.

Szabo, Liz,"Report: Just One Cigarette is Bad," *USA Today*, December 9, 2010,
Taylor, Ian T. *In the Minds of Men.* Minneapolis, Minn.: TFE Publishing, 1991.
__________, "The Ultimate Hoax: Archaeopteryx Lithographica," *Proceedings of the Second International Conference on Creationism, Vol.. II* , Pittsburgh: Creation Science Fellowship, Inc., 1990.
__________, "The Idea of Progress," *The Fifth International Conference on Creationism,* Pittsburgh Creation Science Fellowship, Inc., 2003.
Thomas, Brian, "Fossilized Materials Must Be Young," *Acts & Facts,* June, 2009.
__________, citing D. Criswell, "How Soon Will Jurrasic Park Open? *Acts & Facts.* 35.
__________, "Fossilized Materials Must Be Young," *Acts & Facts,* June, 2009.
Thaxton, Charles B., Walter L. Bradley and **Roger L. Olsen,** *The Mystery of Life's Origin,* New York: Philosophical Library, 1984.
Thompson, Bert, *The Scientific Case for Creation,* Montgomery, Alabama: Apologetic Press, Inc., 2002.
__________,*The Scientific Case for Creation* ,Montgomery, Alabama: Apologetics Press, 2002.
Thompson, Bert and **Brad Harrub,** "15 Answers to John Rennie and *Scientific American's* Nonsense." Montgomery, Alabama: Apologetics Press, 2002.
__________, "*National Geographic* Shoots Itself in the Foot Again," Apologetics Press.Org online r eport, 2004.
Thompson, Ida. *National Audubon Society Field Guide to NorthAmerican Fossils.* New York: Alfred A. Knopf, Inc., 1982.
Twain, Mark, *Life on the Mississippi* (Boston: J.R. Osgood, 1883), 156 as quoted by Brad Harrub, *Reason and Revelation*, May, 2001, 21(5).
USA Today, Editorial Page, " 'Intelligent Design' Smacks of Creation by Another Name," August 9, 2005.
Vail, Tom, *Grand Canyon: A Different View.* Green Forest, Arkansas: Master Books, 003.
Vardiman, Larry, "Are Hurricanes Getting More Destructive?", *Impact #390* El Cajon, California: Institute for Creation Research, December, 2005) iv.
__________, "Greenland Ice Cores," *CreationDigest.com.*, Winter, 2002.
Vardiman, Larry, Andrew A. Snelling and **Eugene F. Chaffin,** Editors, *Radioisotopes and the Age of the Earth,* Vol. 1, 2000 and Vol. II, 2005, published jointly by Institute or Creation Research, E Cajon, California and Creation Research Society, Chino Valley, Arizona.
Vergano, Dan, "Soil Points to Comet Storm That Was Fatal to Mammoth," *USA Today*, January 2, 2009.
__________, "Greenland Glacier Runoff Doubles Over Past Decade," *USA Today*, February 17, 2006.
Von Braun, Wernher, as quote by James Perloff, *Tornado in a Junkyard,* Arlington, Massachusetts: Refuge Books, 1999.
Wachtershauser, Gunter, Letter to Editor, *Science*, 25 October 2002, vol. 298.
Wagner, Dennis, "2009 Annual Report: The Key Darwin and Design Science News Stories of the Year," *Access Research Network,.* December 30, 2009.
__________, "Top Ten Darwin and Design Science News Stories for 2010," ARN.org, /top10, December 21, 2010.
Wallace, Alfred Russell. *The Geographical Distribution of Animals, With a Study of the Relations of the Living and Extinct Faunas as Elucidating Past Changes of the Earth's Surface.* New York: Harper, 1876.
__________, *Darwinism*, London and New York, Macmillan and Co., 1890.
__________, "The Present Position of Darwinism," *Contemporary Review*, August, 1908.
Wang, Samuel and **Ethel R. Nelson** *God and the Ancient Chinese.* Dunlap, Tennessee: Read Books Publisher, 1998.
Ward, Peter D., "Coils of Time," *Discover*, March 1998.
Ward, Peter & Donald Brownlee, *Rare Earth*, Feb 2000.
Wardrop, Murray, "Scientists Draw Squid Using its 150-million-year-old Fossilized Ink," *Telegraph.co.UK*, 19 August, 2009.
Warren, Rick, *The Purpose Driven Life,* Grand Rapids, Michigan: Zondervan, 2002.
__________, "Starbucks Stirs Things Up with a God Quote on Cups," *USA Today*, October 19, 2005.
Washington Post, The, "Genetic Error Causes Rapid-Aging Syndrome,"April 17, 2003.
__________, ***Weekly Edition*, June 13-19, 2005.**
Watson, Lyall, "The Water People," *Science Digest*, 90[5]:44, May, as cited by Eric Lyons and Kyle Butt, *The Dinosaur Delusion*, Montgomery, Alabama: Apologetics Press, Inc., 2008.
Weaver, Daniel C.,"The River of Life," *Discover*, November 1997.
Weiss, Rick. "Defect Tied to Doubling of Risk for Colon Cancer." *The Washington Post*, August 26, 1997.
__________, "Water Scarcity Prompts Scientists to Look Down," *Washington Post*, March10, 2003.
Wells, Jonathan, *Icons of Evolution.* Washington, D.C.: Regnery Publishing, Inc., 2000.
__________, "Survival of the Fittest," *The American Spectator*, December 2000/January 2001.

__________, *Darwinism and Intelligent Design* (Washington, DC: Regnery Publishing,, Inc., 2006).

Wells, Spencer, "From Africa to Astoria by Way of Everywhere," Genographic Project, *National Geographic,* September, 2009.

Wells. W., "Taking Life to Bits," *New Scientist,* 155(2095):30-33, 1997 as cited in "How Simple Can Life Be?, *AnswerwinGenesis.org*, 2009.

Werner, Carl, *Evolution: The Grand Experiment,* Green Forest, Arkansas: New Leaf Press, 2007

West, John Anthony, *The Serpent in the Sky* (New York, NY: Harper and Row, 1984).

Whitcomb ,John C. and **Henry M. Morris,** *The Genesis Flood,* Phillipsburg, New Jersey: Presbyterian and Reformed Publishing Company, 1995.

Wickramasinghe, Chandra, "Threats on Life of Controversial Astronomer," *New Scientist,* January 21, 1982.

Will, George F. "The Gospel of Science." *Newsweek,* November 9, 1998.

Wise, Kurt P., *Faith, Form and Time,* Nashville: Broadman & Holman, Pubs., 2002.

Witham, Larry A., *Where Darwin Meets the Bible.* New York: Oxford University Press, 2002.

Wikedia.org, "Mitochondrial Eve," citing Rohde, DL, Olson, S, Chang, JT, "Modeling the Recent Common Ancestry of all Living Humans," *Nature,* (September, 2004).

Woodward, Thomas, *Doubts About Darwin.* Grand Rapids: Baker Books, 2003.

Wysong, Randy L., *The Creation-Evolution Controversy*East Lansing, Michigan: Inquiry Press, 1976.

Yeld, John ,"Fossil Tracks of Giant Scorpion a World First," *Independent Online 2002*, August 29, 2002.

Yeoman, Barry, "Schweitzer's Dangerous Discovery," *Discover,* April, 2006.

Yockey, Hubert P., *Information Theory and Molecular Biology.* Cambridge: Cambridge University Press, 1992.

Yu, C.P. "The Human Brain Testifies Against Evolution: Confessions of a Neurosurgeon," Internet Website: www.hkam.org.hk/temp/counterevolution, as noted 1-10-200

Zimmer, Carl, "Testing Darwin," *Discover,* February, 2005.

__________, "A New Step in Evolution," internet *Seed Design Series,* posted June 2, 2008.

The author pictured here in 1989, shortly before he initiated investigation of evolution's conjectures as to the origin of life on Planet Earth.

Author

Warren LeRoi Johns, circa 2006

Born in 1929, the year commercial airline services debuted; the first ever Motion Picture Academy Awards presentation; and the onset of the Great Depression, Johns practiced law as a career in California, Maryland, and the District of Columbia until partial retirement in 1992.

Admitted to practice before the United States Supreme Court in 1963, Johns has been a member of the American Association for the Advancement of Science. His 1999 ***Ride to Glory*** used his lawyer's academic perspective targeting evolution's most obvious *"flaws"* and *"holes."*

This title was followed by his 2007 ***Beyond Forever, Genesis File*** in 2009, and 2011's ***Chasing Infinity*** and ***Time Zero.*** Earlier his non-fiction ***Dateline Sunday, U.S.A.***, drew national attention as a legal history documenting blue law confrontation with the U.S. Constitution's first amendment.

A 1958 graduate of the University of Southern California's Law Center, and holder of the Church State Council's 1972 Frank Yost Award and La Sierra University's 1994 "Alumnus of the Year" award, the author's professional resume appears in *Who's Who in American Law*; *Who's Who in America*; and *Who's Who in the World.*

Warren L. Johns, Esq. (ret.)
wj1935@yahoo.com

"...The fool says in his heart, 'There is no God.'" Psalms 14:

More Reading for Inquiring Minds....

Ride to Glory exposes evolution's most notorious flaws
in an action oriented, fictional mock trial staged on a college campus.
First published in 1999, *Ride to Glory's* 2012 edition is now available
in paperback or eBook formats through leading online retailes
along with these other titles by this author.